Balancing Collections Performance and Service Ratings:

Assessing the Impact of Policies at Municipal and Cooperative Utilities

Balancing Collections Performance and Service Ratings:

Assessing the Impact of Policies at Municipal and Cooperative Utilities

Steven E. Seger

American Water Works Association

Copyright © 2006 American Water Works Association
All rights reserved.
Printed in the United States of America.

Disclaimer

The author, contributors, editors, and publisher do not assume responsibility for the validity of the content or any consequences of their use. In no event will AWWA be liable for direct, indirect, special, incidental, or consequential damages arising out of the use of information presented in this book. In particular, AWWA will not be responsible for any costs, including, but not limited to, those incurred as a result of lost revenue. In no event shall AWWA's liability exceed the amount paid for the purchase of this book.

This publication is designed to provide accurate and authoritative information in regard to the subject matter covered. It is sold with the understanding that the author is not engaged in rendering legal, accounting, or other professional service. If legal advice or other expert assistance is required, the services of a competent professional person should be sought.

Project Manager/Senior Technical Editor: Mary Kay Kozyra

Library of Congress Cataloging-in-Publication Data
Seger, Steven E.
  Balancing collections performance & service ratings : assessing the impact of policies at municipal & cooperative utilities / Steven E. Seger.
    p. cm.
  Includes bibliographical references and index.
  ISBN 1-58321-416-X
  1. Public utilities--United States--Rates. 2. Public utilities--United States--Costs. 3. Customer services--United States. 4. Consumer satisfaction--United States. I. Title.

HD2766.S38 2006
363.6068'8--dc22   2005057106

6666 West Quincy Avenue
Denver, CO 80235-3098
303.794.7711

Contents

Acknowledgments, vii

Introduction, ix

1 Linking Policy to Results ---------------------- **1**
Pilot Study, 1
Customer Interactions, 8

2 Pilot Study Results ---------------------------- **11**
Lesson 1: Prioritizing Collections, 11
Lesson 2: Prioritizing Service, 15
Lesson 3: Inactive Timeline, 17
Cross-Tabulation Analysis, 20

3 New Account Requirements ---------------------- **21**
Lesson 4: Application for Service, 21
Lesson 5: Application Fees and Deposits, 22
Lesson 6: Certificates of Deposit, 26
Lesson 7: Credit Scoring, 31
Lesson 8: Credit Databases, 33

4 Current Account Tasks ------------------------- **59**
Lesson 9: Billing Frequency, 35
Lesson 10: Multiple Active Timelines, 39
Lesson 11: Bad Checks, 41
Lesson 12: Payment Extensions and Arrangements, 43
Lesson 13: Late Fees, 46

5 Delinquent Account Actions — — — — — — — — — — — — 51
Lesson 14: Automatic Phone-Dialer Contacts, 51
Lesson 15: Field Collections, 53
Lesson 16: Disconnection/Shut-Off, 56
Lesson 17: Technologies for Disconnection/Reconnection, 60
Lesson 18: Deconstructing Bad Debts, 65
Lesson 19: In-House Versus Outsourcing Collections, 72
Lesson 20: Collections Agency Request for Proposals, 75
Lesson 21: Collections Contract Terms, 77

6 Untapped Potentials — — — — — — — — — — — — — — — 85
Lesson 22: Bankruptcy Processing, 85
Lesson 23: Sweep Accounts, 88
Lesson 24: Selling Debts, 89
Lesson 25: Tracking Accounts Receivable, Collections, and Write-Offs, 93

Appendix A: **45 Credit and Collection Policies, 99**
Appendix B: **Collections Agency Request for Proposals, 103**
Index, 109
About the Author, 112

Acknowledgments

This book is the result of interactions with countless utility professionals, local municipal officials, and directors of cooperatives during courses and consultation projects. Over the years, I taught many of these industry courses through national, regional, and state associations. A special thanks is offered to individuals at a few of these groups:

- Pam Cowen, Cindy McGill, and Kevin Cullather at American Public Power Association
- Pat Mangan and Valerie Taylor at National Rural Electric Cooperative Association
- Barry Fuchs at Northwest Public Power Association
- Jim Wyche and Danette Scudder at Tennessee Valley Public Power Association
- Pat Hyland at Northeast Public Power Association
- and representatives from the many state associations with whom I have worked.

The pilot study that led to this book would not have been possible without the participation of the following groups. I am especially appreciative of the time and support provided by the people named below within these groups:

- Clarksville Department of Electricity: Kenneth Spradlin, Kimberly Satterfield, and Dana Johnson
- Cleveland Utilities: Tom Wheeler, Ken Webb, Brenda Jarrett, and Amy Umiker
- Erwin Utilities: Lee Brown, Tracy Moore, and Becky Renfro
- LaFollette Utilities: Kenny Baird, Cheryl Tidwell, and Linda Bolt
- Middle Tennessee Electric Membership Corporation: C. Frank Jennings and Lisa Carroll

Paris Board of Public Utilities: John Etheridge and Rosemary Brown

Pennyrile Rural Electric Cooperative Corporation: Easton Glover and Sandy Grogan

Trenton Light & Water: Bret Fisher and Jenny Corbin

Tri-County Electric Membership Corporation: Paul Thompson, Glen Hale, and Tammy Dixon

Volunteer Energy Cooperative: William Buchanan, William Schmidt, and Betty James.

In addition, I acknowledge the statistical analysis assistance of Dr. Karl A. Seger, Robin Baker, Dr. Gregory Faulk, and John Gonas. Finally, Jennifer Thomas' editorial skills are much appreciated.

Introduction

The performance of local officials/directors and management of not-for-profit water, wastewater, electric, and natural gas utilities is judged not by a single financial outcome but on multiple factors. Among these factors, two dominant performance indicators are fiscal health and customer satisfaction. At times, actions that are perceived to be positive for a utility's fiscal health (e.g., securing deposits) are viewed as negative in relation to the utility's levels of customer satisfaction. Leaders of not-for-profit utilities are ultimately charged with the responsibility of setting policies that balance or maximize their organization's performance in these two critical areas.

The chapters in this book contain 25 Key Lessons (listed on p. x) concerning the impact of credit and collections policies on revenues and customer satisfaction ratings at not-for-profit utilities, drawing from a number of sources. The conclusions described herein validate some historical views on collections and satisfaction ratings, dispel other myths thought to be obvious or assumed, and contradict a few long-held beliefs. These lessons provide local officials/directors and managers with information of substance that they can make use of while formulating policies and operational procedures.

25 KEY LESSONS

Lesson 1: Prioritizing Collections

Higher collection priority (CP) results are associated with lower write-offs and higher current accounts receivable balances but *not* with lower satisfaction ratings.

Lesson 2: Prioritizing Service

Higher customer service priority (CSP) results are associated with higher write-offs and higher current accounts receivable balances but *not* with higher satisfaction ratings.

Lesson 3: Inactive Timeline

Time is the variable associated most closely with write-offs.

Lesson 4: Application for Service

Increasing available application methods has no effect on write-offs or service ratings.

Lesson 5: Application Fees and Deposits

Higher fees and deposits do not lower write-offs, but they do lower customer satisfaction.

Lesson 6: Certificates of Deposit

Acceptance of certificates of deposit in lieu of cash can eliminate a utility's interest expense on commercial security deposits.

Lesson 7: Credit Scoring

Customizing deposit policy based on customer scoring can reduce write-offs and increase service ratings.

Lesson 8: Credit Databases

Modern utility-specific credit database matching services generate strong financial results.

Lesson 9: Billing Frequency

Maximizing monthly billing frequency serves to minimize bad-debt expenses.

Lesson 10: Multiple Active Timelines
Utilities can positively influence current accounts receivable and minimize bad debt losses by implementing variable timelines.

Lesson 11: Bad Checks
Bad-check policies with limits and appropriate fees minimize the number of checks that utilities must process and the negative impact of these fees on service ratings.

Lesson 12: Payment Extensions and Arrangements
Service ratings neither rise nor fall based on how many extensions or arrangements a utility provides, though authorizing more of them does lead to lower current accounts receivable.

Lesson 13: Late Fees
Nonexistent and inadequate late fees lead to lower average monthly current accounts receivable and increased losses.

Lesson 14: Automatic Phone-Dialer Contacts
The costs generated by notifying customers of a pending shut-off and the numbers of these actions are reduced by the use of phone-dialing systems.

Lesson 15: Field Collections
The overwhelming majority of not-for-profit utilities collecting payments in the field report no incidents of theft or violence.

Lesson 16: Disconnection/Shut-Off
Improper information contained in shut-off notices can lead to legal and, thus, financial risk.

Lesson 17: Technologies for Disconnection/Reconnection
Technology has resulted in the most cost-effective approaches for performing many disconnections and reconnections.

Lesson 18: Deconstructing Bad Debts

Calculating the average daily bad debt is the first step toward deconstructing aggregate losses.

Lesson 19: In-House Versus Outsourcing Collections

Information available primarily via the Internet makes locating debtors a relatively low-cost task for most utility collectors.

Lesson 20: Collections Agency Request for Proposals

The major reason many utilities get inadequate results from collections agencies is the absence of a formal selection process.

Lesson 21: Collections Contract Terms

The second major reason many utilities experience inadequate results from collections agencies is that contractual terms do not support positive results.

Lesson 22: Bankruptcy Processing

The financially negative impact of businesses entering bankruptcy is reduced and even eliminated by utility use of prudent long-term credit policies.

Lesson 23: Sweep Accounts

Leveraging cash assets provides not-for-profit utilities with an additional revenue stream from which to fund ongoing operations.

Lesson 24: Selling Debts

Liquidating uncollected bad debts is a relatively simple and effective business practice for obtaining additional revenues.

Lesson 25: Tracking Accounts Receivable, Collections, and Write-Offs

Monitoring present collections results is the best way to avoid lagging receivables and heightened losses.

SOURCES

The information in this publication was compiled from four primary sources: industry courses, consulting projects, literature review, and primary research. Combined, their contributions provide a broad base, as well as a compelling depth of information regarding collections and service ratings at not-for-profit utilities.

Industry Courses

Credit and collections courses offered to countless local officials/ directors, managers, and employees from not-for-profit utilities in the United States and abroad provided a far-reaching perspective of policies and their effects. Participants recounted the unique and routine approaches their utilities took to address common issues of collections and service. These discussions opened a treasure chest of anecdotal evidence leading to several common themes, or key lessons, that applied equally well from one utility to the next.

Course participants attended sessions scheduled by national, regional, and state utility associations as well as other groups. These associations included, among many other state associations, Tennessee Association of Utility Districts, American Public Power, California Water, National Rural Electric Cooperative, Northeast Public Power, Northwest Public Power, Tennessee Valley Public Power, and Tennessee Gas.

Consulting Projects

In-depth assessments of the structure and operations of several utilities' credit, billing, and collections programs were performed. Also, mail-out and telephone customer surveys, as well as secret shopper programs, were conducted. Utilities in states including California, Georgia, Kentucky, Indiana, Massachusetts, New York, Tennessee, and Texas have undergone these collections and/or service evaluations.

The collections projects in particular were comprehensive and included reviews of several policy and operational areas such as:

- Organization of collection activities and department responsibilities as compared with similar utilities
- Policy reviews: bad checks, disconnects, applications, fees, etc.
- Staffing requirements and utilization
- Management of customer relations in the collections process
- Quantification of losses associated with each deficiency identified
- Performance analysis–cost/benefit of activities and the utility's performance as compared with similar utilities.

Literature Review

A review of the articles published in industry trade journals offered technical information related to collections and service practices. Interviews with industry representatives contained within these articles further described utility experiences with various approaches. In addition, past studies, such as "Why Utility Customers Don't Pay Their Bills" and *Lessons Learned From Mystery Shopper Programs: Assessing the Quality of Service Provided by Public Power Systems* provided detailed information related to highly specific collections and service matters.[1,2] Last, information collected from several industry surveys was reviewed.

In sum, these literature reviews served as the secondary research considerations for this publication. Secondary research is simply the process of analyzing existing data.

Industry courses, consulting projects, and literature reviews made clear that a lot of information was available regarding the collections and service policies not-for-profit utilities chose to implement. For instance, 72% of a survey's respondents might state that they require deposits, but a comprehensive study designed to uncover the impact of policies, such as deposits, on collections performance and service ratings was not found.

Primary Research

It was determined that primary research (information collected firsthand from a sampling of not-for-profit utilities) was needed to uncover existing relationships between policies and collections as well as service performance. Before such a large-scale study could be effectively conducted for all of the electric, water, wastewater, and natural gas not-for-profit utilities in the United States, research protocol dictated that a formal pilot study was necessary.

IN THIS BOOK

Chapter 1 explains how the pilot study was undertaken. Chapter 2 presents the broad findings of the pilot study. Chapters 3 through 5 discuss the 25 lessons as they relate to each of the three sets of customer interaction processes. Finally, chapter 6 covers untapped collections resources.

Notes

1. Michael Kiefer and Ronald Grosse, Why Utility Customers Don't Pay Their Bills, *Public Utilities Fortnightly,* (June 1984):41–44.

2. Steven E. Seger, *Lessons Learned From Mystery Shopper Programs: Assessing the Quality of Service Provided by Public Power Systems,* Washington, DC: American Public Power Association, 2001.

1

Linking Policy to Results

To investigate how various policies are linked to results, a pilot study was undertaken. This study took into consideration the processes involved in customer interactions: new account requirements, current account tasks, and delinquent account actions.

PILOT STUDY

Through this pilot study, several key policies were identified that had considerable influence on a utility's collections performance and service ratings. The pilot study documented various credit requirements and collection approaches from a sampling of not-for-profit utilities and recorded the collections and service performance of each utility studied. Examples of the data include write-offs, accounts receivable aging schedules, collections expenses, overall service ratings, and employee ratings. Then a comprehensive analysis, designed to uncover the impact of each credit and collection activity upon each area of performance, was completed.

The objectives of the pilot study were to:
- Develop and test the adequacy of a research instrument,
- Test the probability that the results were statistically sound, and
- Identify the policies that most likely were influencing collections and service performance.

This chapter covers the pilot study design through testing procedures.

Design

The scientific method of research selected included the following three main stages:

1. Observation or collection of information,
2. Hypothesis development (making educated guesses about the information), and
3. Experimentation of the information testing to prove or disprove traditional notions regarding collections and service ratings.

As with all scientific research, it was important in this pilot study to construct a data-collection tool and procedure that focused on a specific issue rather than on a lot of different topics. When it comes to research, the typical choice is between (a) collecting a lot of information from a few participants or (b) collecting less information from many participants. Also, a rule of thumb regarding information collection is that the more information collected from any single participant, the fewer the number of cases needed to achieve a stable sample estimate.

Because no previous studies seeking to scientifically link policies with performance were found, a detailed tool and time-consuming data collection procedure were selected. Based on the large amount of information that would be necessary to collect from each participant, it logically followed that a small number of participants would provide a stable sample estimate. From this sample, considerations regarding the design of a full study and, as available, conclusions regarding the impact of various policies could be drawn.

The form used to collect information consisted of 90 questions spread across 11 topics. Information was obtained in regard to both residential and business policies. The answers to several questions had multiple parts and, thus, generated more data points, which were used during the analysis phase. The 11 topical areas were:

1. Demographics
2. Application for service
3. Application fees and deposits

4. Meter reading
5. Billing distribution and collection
6. Delinquent payments: notification, fees, payment arrangements
7. Disconnection/shut-off: notification and fees
8. Reconnection: fee, amount due, and deposit
9. Post-disconnection/shut-off collections
10. Credit and collections miscellaneous
11. Customer service

The data collection procedures considered were: survey research, direct measurement, and observation. The decision to use a detailed version of survey research was selected based on the type of pilot study undertaken, as well as the data sought. Among the three methods of survey research (mail-out, telephone, and in-person interviews), face-to-face interviews conducted on-site at each participating utility was the method selected because of its ability to produce the scope and depth of information believed to be necessary to accomplish the objectives of the pilot study.

Participants (Sample)

A population from within the electric, water, wastewater, and natural gas industries was identified as a prudent study group to represent the industries as a whole because of three predominant, homogenous characteristics:

1. The industries targeted in the study all functioned as creditors providing their services prior to receiving payments.
2. The technologies, policies, and business practices used to facilitate collections were nearly identical regardless of the actual product (e.g., electricity) provided.
3. They all functioned as not-for-profit entities.

After investigation, a group, or sample, of utilities was identified that strongly represented the industries as a whole. The Tennessee Valley Authority (TVA) is a quasi-governmental organization that, among other enterprises, serves as the primary generation and transmission utility to distributors in seven southeastern states. These 158 distributors all belong to the Tennessee Valley Public Power Association (TVPPA) and provide one or more of the following services: water, wastewater, electricity, and natural gas. Some of these utilities also provide ancillary services, such as long-distance service. This group was selected for the pilot study sample because the distributors all function as creditors facilitating collections using the same or similar approaches while functioning as not-for-profit local utilities.

Based on the three predominant, homogenous characteristics of the pilot study population, it was projected that a sample consisting of roughly 5% would be necessary to obtain an appropriate quantity of data for analysis. Accordingly, 10 utilities (6.33%) from among the 158 in the total population were requested, and agreed, to participate in the pilot study. The number of customers of the participating utilities is noted in Figure 1-1.

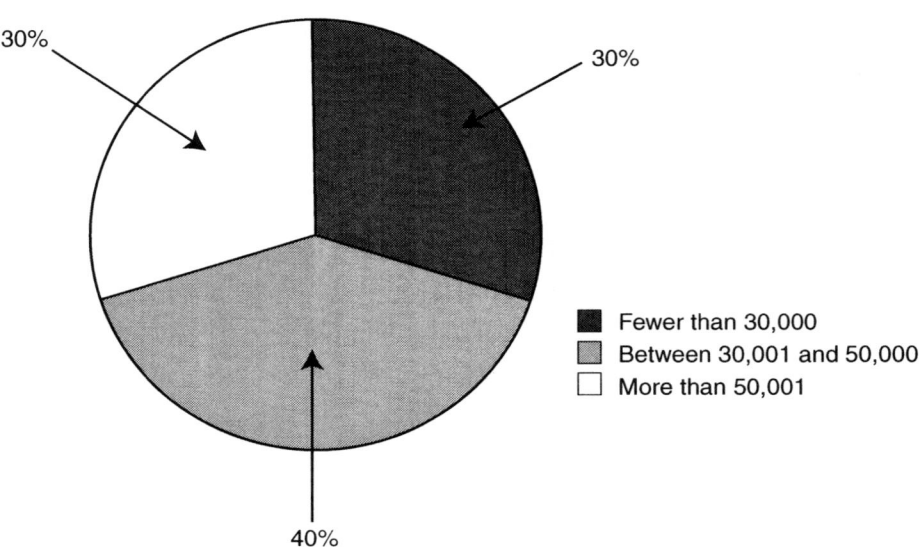

Figure 1-1. Number of Customers of Participating Facilities

Data Collection

Once the pilot study design was complete and the participants were identified, data collection and organization began. The survey questionnaire was completed during a series of interviews with participating utility staff members. In addition, many items related to credit, billing, collections, and service were collected. Following are some of these items:

Annual budget	Extension agreements
Service policies	Monthly bill format
Annual report	Shut-off reports
Accounts receivable aging schedule	New customer packets
Customer satisfaction surveys	Service order systems
Collection agency data	Late and shut-off notices
Bankruptcy data	Commercial deposit forms
Application for service	Customer letters
Write-off reports	Billing system information
Collection agency contracts	CIS (customer information system) screens

The process of organizing the information or preparing it for testing involved the following four steps.

Step 1

First the data were entered into an Excel spreadsheet. During this data entry process, some of the 90 questions on the survey were subdivided, producing a total of 105 questions. Combined, these 105 questions answered by the 10 utilities produced 2,590 facts for analysis. For instance, the question in Table 1-1 pertaining to a utility's requirements for signed applications offered 10 facts.

Table 1-1. Question Pertaining to a Utility's Requirements for Signed Applications

Do residential customers have to sign an application for service?										
Utility	1	2	3	4	5	6	7	8	9	10
Answer	Yes	Yes	Yes	Yes	No	Yes	No	No	Yes	Yes

Step 2

Forty-five policies that seemed to be the most likely factors influencing collections and service performance were identified (see appendix A). These policies served as the independent variables that would be used during the testing. Independent variables affect dependent variables, and a policy requiring customers to sign an application for service was one independent variable that was tested for its effect on write-offs and service ratings (i.e., dependent variables).

Step 3

Likewise, for testing purposes, several collections outcomes and service ratings were selected as performance indicators. These items served as the dependent variables.

Step 4

Last, the information was structured in a format conducive for testing. The median point of the 45 policies was identified. Answers above the median point indicated that a utility had a higher collection priority (CP) regarding that policy than other utilities with answers below the median point. In contrast, answers below the median had a higher customer service priority (CSP). Finally, answers that were in a yes-or-no format were labeled as CP or CSP based on the circumstances involved (see Table 1-2). Requiring a person to sign an application is an example of a CP, and the absence of such a requirement is considered a CSP.

Table 1-2. Yes/No Answers Labeled as CP or CSP

	Policy Scales		
Utility	CP	At Median	CSP
1	20	9	16
2	12	16	17
3	13	14	18
4	24	13	8
5	14	14	17
6	15	14	16
7	14	15	16
8	22	13	10
9	16	15	14
10	23	8	14

Hypotheses

At this point, the facts regarding policies and performance outcomes were ready for the next phase in the study: hypothesis development. Several historical notions regarding the effects of credit and collections policies on financial and service performance served as the first set of hypotheses. Once a series of tests was performed, additional lessons became obvious. The most prominent hypotheses or common beliefs originally tested are as follows:

1. Effects of prioritizing collections or maintaining more conservative collections policies:
   a. Utilities with CP scores above the sample median will have lower write-offs.
   b. Utilities with CP scores above the sample median will have a higher average current accounts receivable percentage.
   c. Utilities with CP scores above the sample median will have lower service ratings.
2. Effects of prioritizing customer service or maintaining more liberal collections policies:

a. Utilities with CSP scores above the sample median will have higher write-offs.
   b. Utilities with CSP scores above the sample median will have a lower average current accounts receivable percentage.
   c. Utilities with CSP scores above the sample median will have higher service ratings.
3. Effects of the length of time between finalizing an account and initiating collections actions for non-payment:
   a. Utilities that more quickly initiate collection actions will have lower write-offs.
   b. Utilities that initiate collection actions more quickly will have lower service ratings.

Testing

"Analyse-It General Statistical Module" software was first used to perform a descriptive analysis of the data sets. The data were found to be non-parametric, which meant that the data were not disbursed evenly around a common mean. Next, the Wilcoxon signed-ranks test was utilized. This test was conducted to establish the probability that conclusions drawn from the data would be significant (i.e., correct).

Last, a series of cross-tabulated contingency tables were designed to examine the relationship between variables. The objective of these tests was to establish the degree of influence that one variable (e.g., deposits) may have over another variable (e.g., collections performance).

In sum, testing was done to understand the type of data collected, to determine whether conclusions could be reliably drawn from the data, and to learn what lessons the data made available for policy makers and management of not-for-profit utilities.

CUSTOMER INTERACTIONS

It has been observed that not-for-profit utilities basically follow three sets of processes when it comes to customer interactions: new account requirements, current account tasks, and delinquent account actions. Key lessons regarding new account requirements, such as methods made available to apply for service and deposit requirements, demonstrated expected as well as some surprising results. For example, utilities with higher deposits had lower customer satisfaction scores, but they did not have lower write-offs.

Current account tasks consist mainly of reading meters, billing for usage, and processing payments. In general, these tasks are carried out in fairly consistent ways across the industry, although a few important lessons related to billing and payment processing are included in this book.

Delinquent account actions are those processes that occur when a bill is not paid by its due date. Setting aside the various types of late and disconnect/shut-off notices used, the flowchart in Figure 1-2 illustrates typical steps that take place when a bill goes unpaid. As depicted, local officials/directors and management of not-for-profit utilities face various policy decisions when considering the most effective ways to pursue collections. A few of the policy matters that leaders must address are: offering extensions, using staff to perform collections, outsourcing collections to agencies, and managing write-off accounts.

We present the results of the hypotheses tested against the pilot study data, as well as several other key lessons drawn from industry courses, consulting projects, and secondary research, in the next chapter.

BALANCING COLLECTIONS PERFORMANCE AND SERVICE RATINGS

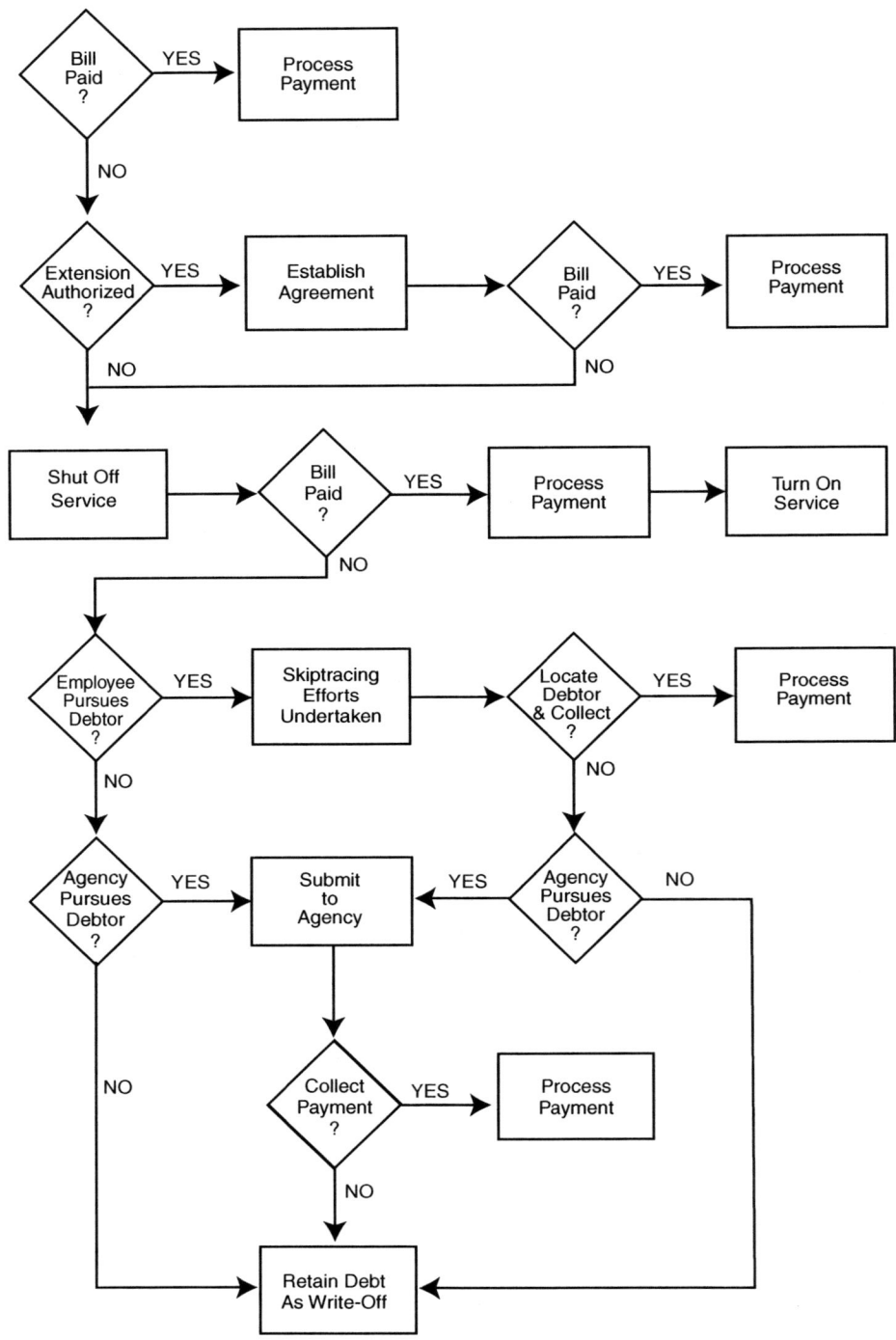

Figure 1-2. Typical Steps When a Bill Goes Unpaid

2

Pilot Study Results

The pilot study had three objectives:
1. To develop and test the adequacy of a research instrument. As discussed in chapter 1, the instrument developed for data collection was successful in identifying 45 policies that demonstrated varying degrees of influence on collections and service performance.
2. To test the probability that the results were statistically sound.
3. To identify policies that most likely influence collections and service performance.

LESSON 1: PRIORITIZING COLLECTIONS

Higher collection priority (CP) results are associated with lower write-offs and higher current accounts receivable balances but not *with lower satisfaction ratings.*

The CP for each utility is listed in Tables 2-1, 2-2, and 2-3. In addition, the tables provide the corresponding write-off percentage, current accounts receivable, or satisfaction ratings for each utility. As discussed previously, CPs were determined by identifying the median data point for each of the 45 policies. Each CP figure in the tables represents the number of instances in which a utility scored above the median data point for the group. For instance, utility 1 had 20 credit and collections policies that were more strict than the group's median.

BALANCING COLLECTIONS PERFORMANCE AND SERVICE RATINGS

Table 2-1. Effect of Collections Priority (CP) on Write-Offs (WO)

Utility	CP	WO (%)
1	20	0.20
2	12	0.21
3	13	0.10
4	24	0.39
5	14	0.16
6	15	0.48
7	14	0.19
8	22	0.38
9	16	0.30
10	23	0.20

Table 2-2. Effect of Collections Priority (CP) on Current Accounts Receivable (A/R)

Utility	CP	A/R
1	20	0.95
2	12	0.94
3	13	0.91
4	24	0.87
5	14	0.83
6	15	0.96
7	14	0.82
8	22	0.84
9	16	0.94
10	23	0.93

PILOT STUDY RESULTS

Table 2-3. Effect of Collections Priority (CP) on Satisfaction Ratings

Utility	CP	Satisfaction Rating (%)
1	20	44
2	12	66
3	13	75
4	24	54
5	14	62
6	15	64
7	14	40
8	22	56
9	16	63
10	23	64

Table 2-4. Results of Wilcoxon Signed-Ranks Test (n = number)

Difference Between Pairs	n	Rank Sum	Mean Rank
Positive	10	55.0	5.50
Negative	0	0.0	—
Zero	0		
Difference between medians	89.620		
95.1% confidence interval	59.810 to 109.655 (exact)		
Wilcoxon's W statistic	55		
p-value	0.0020 (exact)		

The Wilcoxon signed-ranks test was performed on each of these data tables to determine if the data were statistically sound and whether the hypotheses from chapter 1 should be accepted. For example, the first hypothesis to be tested was that utilities with CP scores above the sample median would have lower write-offs. The results of the Wilcoxon signed-ranks test for this hypothesis are given in Table 2-4.

For this pilot study, in which the p-value was found to be less than .05, or 5%, the data is sound. The p-value regarding CP and write-offs was

0.0020, which is less than 0.05; therefore, the data were found to be reliable for analysis. The next step was to determine what the data revealed about these items.

Wilcoxon's W statistic communicates whether an initial guess or hypothesis regarding data should be accepted. The calculated W statistic is compared to critical values appearing in the statistical resource, *Some Rapid Approximate Statistical Procedures*.[1] If a calculated value exceeds the reported critical value, the hypothesis is accepted.

The calculated W statistic from the data was 55, and the critical value located in the statistical resource was 10. Accordingly, the hypothesis that utilities with CP scores above the sample median were associated with lower write-offs was accepted. In other words, stricter credit and collections policies were associated with lower write-offs.

The W statistic and p-value results of the Wilcoxon signed-ranks test for the second hypothesis, "higher CP scores are associated with higher current accounts receivable percentages," are provided in Table 2-5. As with write-offs, this information communicates the existence of sound data (i.e., p-value) that reveals a high degree of association between a collections priority and higher current accounts receivable.

The last test looking into a possible relationship between a utility's CP and its service rating was conclusive in an opposite way from write-offs and receivables. The data were deemed to be statistically sound because the calculated p-value was less than 0.05. But the calculated W statistic was less than the critical statistic (0<10). As such, no association between a strong collections priority and low satisfaction ratings could be shown. One item was simply not related to the other one.

Table 2-5. Effect of Collections Priority on Current Accounts Receivable

	Calculated	Critical/Parameter
W statistic	55	10
p-value	0.0020	0.05

LESSON 2: PRIORITIZING SERVICE

Higher customer service priority (CSP) results are associated with higher write-offs and higher current accounts receivable balances but not *with higher satisfaction ratings.*

The three hypotheses concerning the effects of prioritizing customer service or maintaining looser collections policies are restated here.

1. Utilities with CSP scores above the sample median will have higher write-offs.
2. Utilities with CSP scores above the sample median will have a lower average current accounts receivable percentage.
3. Utilities with CSP scores above the sample median will have higher service ratings.

The CSP for each utility is listed in Tables 2-6, 2-7, and 2-8. Again, the corresponding write-off percentage, current accounts receivable, and satisfaction rating for each utility are also provided.

Table 2-6. Effect of Customer Service Priority (CSP) on Write-Offs (WO)

Utility	CSP	WO (%)
1	16	0.20
2	17	0.21
3	18	0.10
4	8	0.39
5	17	0.16
6	16	0.48
7	16	0.19
8	10	0.38
9	14	0.30
10	14	0.20

Table 2-7. Effect of Customer Service Priority (CSP) on Current Accounts Receivable (A/R)

Utility	CSP	A/R
1	16	0.95
2	17	0.94
3	18	0.91
4	8	0.87
5	17	0.83
6	16	0.96
7	16	0.82
8	10	0.84
9	14	0.94
10	14	0.93

Table 2-8. Effect of Customer Service Priority (CSP) on Satisfaction Rating

Utility	CSP	Satisfaction Rating (%)
1	16	44
2	17	66
3	18	75
4	8	54
5	17	62
6	16	64
7	16	40
8	10	56
9	14	63
10	14	64

The p-value and Wilcoxon signed-ranks test for both write-offs and current accounts receivable demonstrated that both sets of data were statistically sound and that they were strongly associated. Higher CSP scores were related to higher write-offs. Obviously, this is not good. In contrast, higher CSP scores were associated with higher current accounts receivable. People who were happier with their utilities paid their bills on time more often than individuals who were not as pleased.

The CSP on the p-value for satisfaction rating calculated with the Wilcoxon signed-ranks test was less than the 0.05 parameter. Therefore, the data offered reliable information. But the calculated W statistic was found to be less than the critical value, so no association between loose credit and collections policies and measured service ratings could be established. This could mean that offering more lenient policies does not necessarily buy a not-for-profit utility goodwill (i.e., higher service ratings) from its customers.

LESSON 3: INACTIVE TIMELINE

Time is the variable most closely associated with write-offs.

The statistical evidence was resounding concerning the degree of association between the period of time a utility waits from finalizing an account until it initiates active collections. This period of time was referred to as the "inactive timeline" in the pilot study. The p-value and Wilcoxon signed-ranks test supported the reliability of the data and its conclusions regarding the presence of a strong relationship. Table 2-9 and Figure 2-1 illustrate these results.

Utilities that waited longer before taking steps to begin the collections process after accounts had been finalized paid a price for this delay with higher write-offs. Figure 2-1 illustrates this linear relationship between time and write-offs. Ultimately, more time equals higher losses.

Table 2-9. Results of Wilcoxon Signed-Ranks Test

Utility	Inactive Timeline (Days)	Write-Off (%)
1	60	0.20
2	100	0.21
3	30	0.10
4	120	0.39
5	60	0.16
6	120	0.48
7	60	0.19
8	90	0.38
9	120	0.30
10	80	0.20

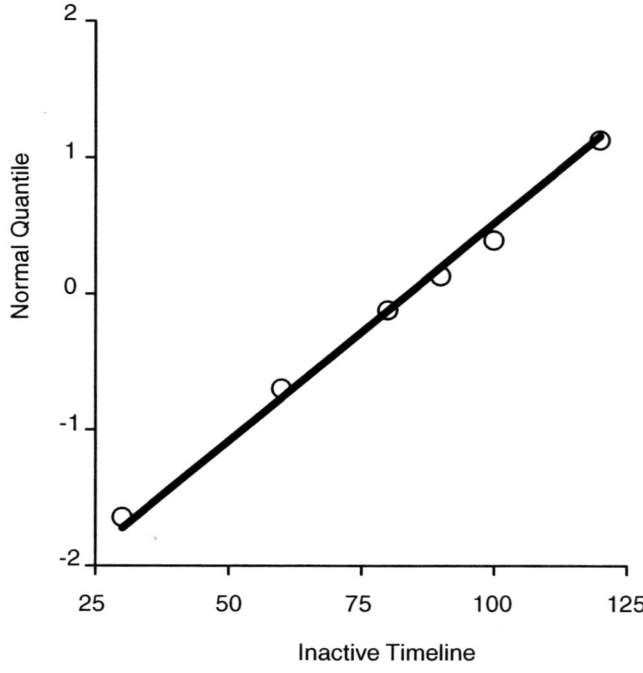

Figure 2-1. Inactive Timeline and Write-Offs

The last hypothesis tested was that utilities that more quickly initiate collections actions will have lower service ratings. The Wilcoxon signed-ranks test results are provided in Table 2-10.

The p-value is only slightly below the established threshold (0.0488 versus 0.05). Therefore, for purposes of the pilot study, the possibility of error is considered to be too great to draw conclusions regarding an association between inactive collections and service ratings. But subjective consideration of the hypothesis does lead to a reasonable consideration.

By definition, the inactive collection timeline addressed only the customers who had left a utility. They were not present in the vast majority of cases, if not all of them, to take part in the collection of satisfaction ratings. Only the few individuals who had gone through this collection experience with a utility and then moved back to the same geographical area and received services from the same utility again would have had their service opinions influenced by this situation. The hypothetical constraints of this scenario make it a nearly nonexistent possibility.

Table 2-10. Results of Wilcoxon Signed-Ranks Test

Difference Between Pairs	n	Rank Sum	Mean Rank
Positive	8	47.0	5.88
Negative	2	8.0	4.00
Zero	0		
Difference between medians	27.000		
95.1% confidence interval	5.500 to 45.500 (exact)		
Wilcoxon's W statistic	47		
p-value	0.0488 (exact tables used, 40% ties)		

CROSS-TABULATION ANALYSIS

The first three lessons of this report substantiated or refuted the existence of a relationship between credit and collections policies and three broad items: write-off percentages, current accounts receivable, and service ratings. Beliefs commonly held throughout the industry were tested against the results from a pilot study. The data supported some of these hypotheses and did not support others.

The next step in the process of identifying clear, supported lessons for policy makers and management involved consideration of information obtained during industry courses, consulting projects, literature reviews, and a more detailed analysis of results of the pilot study. A series of cross-tabulated contingency tables was constructed for this purpose. These tables went beyond the general conclusions made possible from the collection priority and customer service priority rankings. Through these tables, individual policies, such as security deposit amounts, could be tested for any impact they might have on collections performance and service ratings.

Lessons 4 through 25 are based on information collected from numerous not-for-profit utilities across the United States. As with lessons 1 through 3, some of these conclusions may be expected, but several others may come as surprises. The data and examples provided throughout the balance of this book explain the basis for each one of the key lessons.

Notes

1. Frank Wilcoxon and Roberta A. Wilcox, *Some Rapid Approximate Statistical Procedures*, Pearl River, NY: Lederle Laboratories, 1964. Reproduced with the permission of the American Cyanamid Company.

3

New Account Requirements

As creditors, utilities impose a range of requirements on applicants for their services. These requirements typically are meant to assess an applicant's ability to pay for the utility's services, as well as to collect information that could be used for collection purposes, as needed. The number and types of requirements can vary greatly from one utility to the next. In this chapter traditional notions regarding the value of this information and the impact of imposing certain application requirements are evaluated.

LESSON 4: APPLICATION FOR SERVICE

An increase in available application methods has no effect on write-offs or service ratings.

A cross-tabulation of write-offs by customers, who may apply by phone, showed no evidence of increasing losses in cases in which phone applications were permitted. The percent of utilities permitting applications via phone with write-offs <.20, .20 –.30, or >.30 matched the percent not authorizing phone applications in each write-off category. These study results match similar findings collected during industry seminars and consulting projects.

In addition, the percent of utilities offering two or fewer methods to apply had the same general overall opinions of customer service as utilities offering three or more options (see Table 3-1). In-person, over the phone, via Internet, and by fax are examples of methods by which

Table 3-1. Overall Opinion of Service by Methods to Apply for Service

Requirement	Utility Dispersion (%)	Customer Service Score (%)			
		<50	50–60	61–70	>70
2 or fewer	40	25	25	50	0
3 or more	60	17	17	50	17
	100				

applicants could apply at the participating utilities. Past customer survey results mirrored these findings.

Examples of the applications used by two study participants during the period of time evaluated are given in Figure 3-1. They illustrate the quantity and types of information typically collected by the participating utilities.

LESSON 5: APPLICATION FEES AND DEPOSITS

Higher fees and deposits do not lower write-offs, but they do lower customer satisfaction.

Fees and deposits collected during the application process are influenced by several factors. Forces that shape these monetary requirements include, among others, state laws, local ordinances, public policy, and management priorities. Table 3-2 lists the dollar amount and acceptable forms of deposits for several not-for-profit utilities. This information was collected from participants during an industry course.

Study results concerning the perceived impact of fees and deposits on write-offs contradicted long-held beliefs. Utilities with a collection priority (CP) or higher minimum deposit for owners and renters had basically the same write-offs as utilities with a customer service priority (CSP) or lower minimum deposit (see Table 3-3). In addition, the acceptance of letters of credit with no monetary value from former utilities showed neither a substantial positive nor negative effect on write-offs. Of the utilities accepting letters of credit, 75% had write-offs

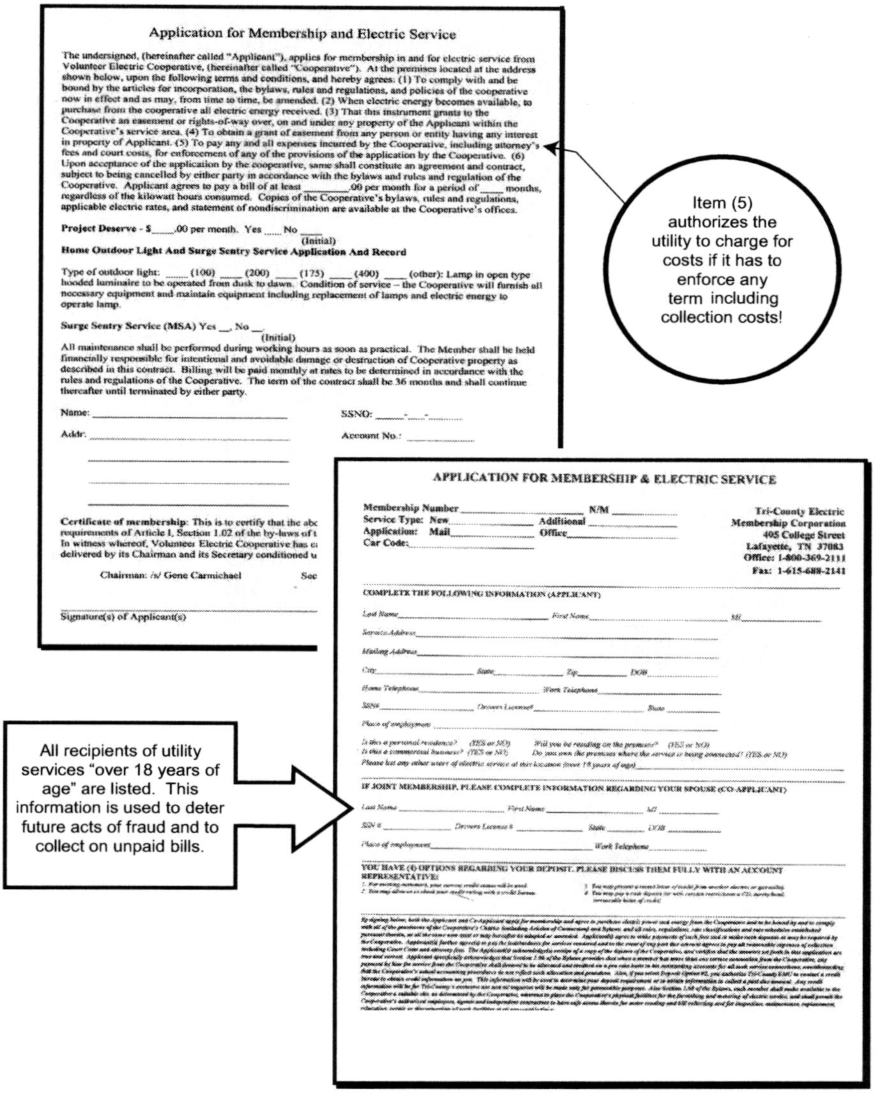

NEW ACCOUNT REQUIREMENTS

Figure 3-1. Sample Applications for New Accounts

greater than or equal to .20. Similarly, 67% of utilities refusing these letters documented write-offs greater than or equal to .20.

The evidence collected regarding the impact of fees and deposits on service ratings revealed a clear inverse relationship. Utilities with higher

Table 3-2. Deposit Policies

Utility	Residential	Commercial
1	Credit scoring with maximum $200 and minimum $100. LC and GA. Retain deposit for life of service.	Two times highest bill. LC or SB. Retain deposit for life of service.
2	$175 renters and $75 owners. Retain deposit for 4 years.	Two times highest bill. CD or SB. Retain deposit for life of service.
3	$150 renters and $100 owners. Retain deposit for 1 year.	Two times average bill from previous 12-month period. LC. Retain deposit for life of service.
4	$150 or $75 with an LC for both renters and owners. Retain deposit for life of service.	Two times highest bill. CD, SB, or LC. Retain deposit for life of service.
5	$150 renters and $100 owners. Retain deposit for 2 years.	Two times highest bill. LC or SB. Retain deposit for life of service.

CD = Certificate of deposit
GA = Guarantor or cosigner agreement
LC = Letter of credit
SB = Surety bond

Table 3-3. Write-Off by Minimum Deposit for Renters

Requirement	Utility Dispersion (%)	Write-Off (%) < .20	.21–.30	> .30
$0–100	50	40	40	20
$101 and up	50	20	40	40
	100			

new customer fees and transfer fees, for instance, had lower service ratings. Figure 3-2 depicts this relationship for new customer fees.

The amount of deposits and duration they were retained by the study group of utilities showed the same inverse relationship as did fees. Examples of this situation are provided in Table 3-4. Read these as, for

NEW ACCOUNT REQUIREMENTS

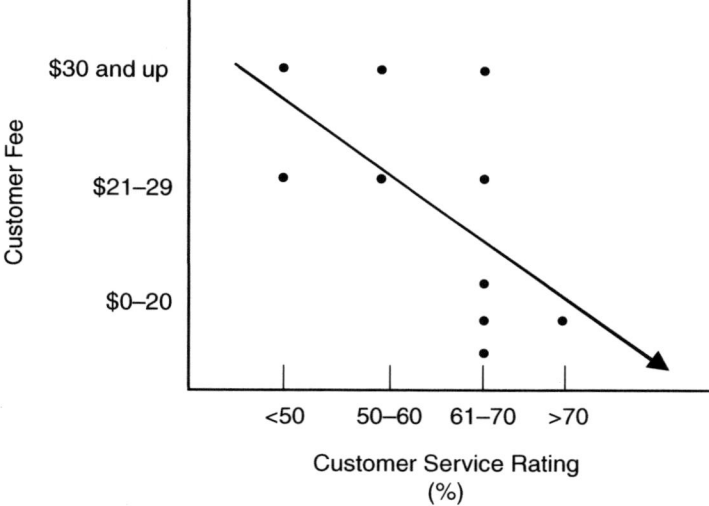

Figure 3-2. Relationship Between Customer Fee and Customer Service Rating

Table 3-4. Examples of Minimum Deposits Related to Service Ratings

Utilities	Minimum Deposit for Owners	Service Ratings
83%	$0–100	>60%
60%	$100 and up	<61%
Utilities	Minimum Deposit for Renters	Service Ratings
80%	$0–100	>60%
60%	$100 and up	<61%
Utilities	Years Deposit Kept	Service Ratings
80%	2 or fewer	>60%
60%	More than 2	<61%

example: "Eighty-three percent of 'Utilities' with 'Minimum Deposit for Owners' of $0–100 had 'Service Ratings' of >60%."

Three credit management tools used by members of the study group providing water and other utility services are shown in Figures 3-3, 3-4, and 3-5. These tools are used to achieve the objective of all creditor organizations: to balance risk with security while considering the impact on service. Deposit insurance is offered through local utilities by the providers of their commodities (i.e., electricity, water, natural gas); a utility service guaranty bond or surety bond comes from an insurance company; and an irrevocable letter of credit is issued by a bank. All three tools provide needed security to utilities while minimizing, or at times eliminating, the financial burden to high-use, business customers.

LESSON 6: CERTIFICATES OF DEPOSIT

Acceptance of certificates of deposit in lieu of cash can eliminate a utility's interest expense on commercial security deposits.

Countless not-for-profit utilities pay interest on the security deposits they collect. Whether interest is paid because of state statutes, regulatory requirements, contractual requirements, or simply as a matter of public policy, this practice ultimately costs utilities a great deal of money. Fortunately, an option allows not-for-profit utilities to avoid paying interest to their highest depositors—commercial accounts—even when this practice is mandated. Unfortunately, many, if not most, utilities do not take advantage of this resource.

Certificates of deposit (CDs) are investment instruments made available from financial institutions such as local banks. CDs typically are "purchased" by individuals or businesses and pay interest just like any other investment. To avoid paying interest out of a utility's general funds, some utilities accept CDs in lieu of cash deposits from their commercial customers. Then the financial institution actually pays the interest to the customer. An example of how these instruments are structured is a one-year CD with an automatic rollover effect.

Exhibit "A"

Enhanced Security Deposit Program

Credit Application & Nomination Form

Credit Application Section (to be completed by Applicant)

APPLICANT: ____________________

(List all name(s) under which the nominated account will contract for or be billed for power and energy.)

Applicant Address: ____________________

(Include street address, city, state, & zip code.)

Telephone: ____________________

DUNS No. ____________________ ☐ **Additional financial information is attached.**

(Providing additional financial information may facilitate faster processing of this application.)

By the signature of its duly authorized representative and as an application for participation in the Enhanced Security Deposit Program (Program) being jointly conducted by Distributor and its wholesale power supplier, the Tennessee Valley Authority (TVA), Applicant hereby:

1. represents and warrants that the above information and any attached financial information is true and correct;
2. authorizes such inquiries as are considered necessary by the insurer or underwriter used by TVA for the Program (hereinafter "Underwriter") to obtain credit information regarding Applicant and authorizes banks, creditors, or other entities having such credit information to release such information to the Underwriter;
3. agrees to furnish such financial information (including, without limitation, balance sheets, income statements, cash flow statements and notes to financial statements) as might be requested by the Underwriter in connection with its review of this application; and
4. agrees, if nominated and approved for participation in the Program on the basis of credit insurance obtained by TVA (hereinafter "Insurance Coverage"):
   (a) to furnish such financial information as might from time to time be requested by the Underwriter or TVA, and
   (b) authorizes such inquiries as are considered necessary by the Underwriter or TVA to from time to time obtain credit information regarding Applicant and authorizes banks, creditors, or other entities having such credit information to release such information to the Underwriter or TVA, and
   (c) to immediately substitute or provide alternative security, in a form and amount that TVA deems acceptable, upon any lapse in the Insurance Coverage for any reason, including, without limitation, the Underwriter or TVA (i) canceling the coverage for any reason or (ii) not renewing the coverage for any reason;
5. agrees that as a part of such alternative security arrangements Distributor may employ weekly or other extraordinary meter-reading or billing arrangements calculated to facilitate such alternative security arrangements;
6. agrees that any failure to substitute or provide alternative security will be deemed to be a breach of the Applicant's power supply arrangements with the Distributor;
7. agrees that if Applicant does not fully remedy such a breach by the date falling 30 days after the date when it first arises, Distributor shall have the right, upon 15 days' written notice, to discontinue the supply of power to Applicant and to refuse to resume delivery until Applicant substitutes or provides alternative security to fully remedy the breach; and
8. agrees that any discontinuance of supply in accordance with 7 above shall not relieve Applicant of liability for minimum monthly charges or payment of past due amounts.

Signature: ____________________ Title: ____________________

Nomination Section (to be completed by Distributor)

DISTRIBUTOR: ____________________

(List all name(s) under which power and energy will be sold or billed by Distributor.)

Requested Insurance Coverage Amount: $ ___

Distributor Contact Name: ____________________

E-mail Address: ___ **Telephone:** ___ **Fax:** ____________________

By the signature of its duly authorized representative and in accordance with and subject to the Enhanced Security Program Agreement and Guidelines, Distributor hereby:

1. nominates the Applicant for participation in the Enhanced Security Deposit Program, and
2. represents and warrants that (a) said nominated account is a new or existing commercial or industrial customer from which Distributor would otherwise require a cash deposit or other substantially equivalent form of security, (b) the Requested Insurance Coverage Amount set forth above reflects the amount of such deposit or security that Distributor would otherwise so require, and (c) that the Credit Application section of this form has been completed by a duly authorized representative of the Applicant.

Signature: ____________________ Title: ____________________

Figure 3-3. Example: Deposit Insurance

```
            UTILITY SERVICE GUARANTY BOND
     KNOW ALL MEN BY THESE PRESENTS, That _____
Principal, hereinafter called Principal, and _____
_____
As Surety, hereinafter call Surety, are held and firmly bound unto
THE LENOIR CITY UTILITIES BOARD; LENOIR CITY, TENNESSEE, as Obligee,
hereinafter called Obligee, in the aggregate sum of
_____, for the payment of which sum will and truly
to be made, we the Principal and Surety above named bind ourselves, our
heirs, executors, administrators and successors, jointly and severally
by these presents.
     WHEREAS, The Principal has requested, and the Obligee has agreed
to furnish utility service to the Principal pursuant to the rates,
rules and regulations for the Company promulgated by proper regulatory
authority having jurisdiction; and
     WHEREAS, The Obligee is willing to accept this Bond in lieu of
securing a cash deposit to be made by the Principal to secure payment
for the services to be furnished.
     NOW, THEREFORE, If the said Principal shall pay or cause to be
paid all bills, statements or charged for any services furnished or
rendered from and after _____, until date of disconnection,
then and in that event, this bond and all obligations hereunder shall
terminate and cease, otherwise, shall remain in full force and effect.
     It is also understood and agreed that Surety may cancel this bond
by written notice served by registered mail upon the LENOIR CITY UTILITIES
BOARD specifying the effective date of said cancellation, which in no
event shall be less than sixty (60) days after the date borne by Surety's
receipt.  But the Surety shall, nevertheless, remain liable for any and
all accrued indebtedness of the Principal to the Obligee incurred prior
to the proposed termination date.
     IN WITNESS WHEREOF, The above parties have executed this instrument
under their several seals, the name and corporate seal of each corporate
party being hereto affixed, and these presents duly signed by its under-
signed representative pursuant to authority of its governing body, this
_____ day of _____, 19__.


                                    _____
                                              PRINCIPAL
                              BY:_____


                                    _____
                                              SURETY
                              BY:_____
                                           ATTORNEY-IN-FACT
```

Figure 3-4. Example: Utility Service Guaranty Bond

How much interest do not-for-profit water, wastewater, electric, and natural gas utilities pay on security deposits to commercial customers each year? A small-scale example serves to provide an indication of the answer to this question. Of the study-group utilities, 60% paid interest

NEW ACCOUNT REQUIREMENTS

IRREVOCABLE LETTER OF CREDIT

LAFOLLETTE UTILITIES
LaFollette, Tennessee 37766

RE: (Customer's name or account name and account address)

This will advise that we at the _____ Bank of _____
are happy to extend to _____ of LaFollette, Tennessee, an
irrevocable letter of credit for the benefit of the LaFollette Utilities Board of LaFollette,
Tennessee, in the sum of $_____. This letter of credit is irrevocable for a period
of one (1) year from the date hereof and thereafter until revoked by this Bank upon not less than
thirty (30) days written notice extended from this Bank to you or Your Company. Our Bank
agrees to issue you a cashier's check for any amount up to the aggregate of $_____ on
any one or more of the following occasions:

   1.   Upon notice that _____ has failed to pay the
LaFollette Utilities Board a bill for utility service within the thirty (30) days of billing. The
Bank will within ten (10) days issue its cashier's check directly to the LaFollette Utilities Board
by standard U.S. mail.

   2.   Upon notice that _____ is insolvent or in
receivership, bankruptcy, or any other re-arrangement of its business affairs which indicates an
inability to meet its financial obligations when due.

   This Bank gladly issues this irrevocable letter of credit in behalf of _____,
and stands ready to be of any assistance if possible, either in the guaranteeing of the above
payments or in granting any references in behalf of said Corporation.

**THIS LETTER OF CREDIT IS A SPECIMEN ONLY. ACTUAL LETTER OF CREDIT
MUST BE TYPED ON ISSUING BANK'S STATIONERY.**

Figure 3-5. Example: Irrevocable Letter of Credit

on deposits to commercial accounts. This is about the same percentage of utilities paying interest that have been identified during industry courses and consulting projects. Based on this figure, the interest costs are estimated in the example shown in Figure 3-6.

A broader example using this same methodology to estimate the interest expense paid by drinking water and wastewater utilities uncovers a more startling figure. Only considering revenues from sales to ultimate

```
  $605,042,096   Revenues of utilities paying interest
         × 12    Months in a year
  _____
    50,420,175   Average revenues per month
         × .45   Commercial portion of revenues per month
  _____
    22,689,078
         × 2     2 months typical commercial deposit amount
  _____
    45,378,157
         × .8    80% of commercial deposits paid in cash
  _____
    36,302,526   Dollar amount of commercial cash deposits
         × .03   3% average interest rate paid
  _____
  $  1,089,075   Interest expense per year
```

Figure 3-6. Example of Interest Cost Estimates on CDs

customers in one year, these utilities paid approximately $57,060,000 in interest.[1] Regardless of how commercial deposits have been addressed in the past, the option to use CDs as a financial tool to secure utility accounts can have the effect of eliminating a tremendous burden on the general revenues of not-for-profit utilities.

LESSON 7: CREDIT SCORING

Customizing deposit policy based on customer scoring can reduce write-offs and increase service ratings.

Credit scoring is the process of assigning a numerical value to an applicant's credit worth or probability of default. The lower an applicant's score, the higher the applicant's probability of default, thus, the higher the required deposit, typically. Increasingly, not-for-profit utilities are utilizing credit-scoring services offered by companies such as Equifax, Experian, and TransUnion in an effort to reduce their bad-debt expenses.

Deposit amounts assigned to each range of credit scores should be customized for each utility to maximize the benefit that can be received from these services. As reported in lesson 5, the existence of increasing deposits demonstrated no effect on write-offs but did have a negative impact on service ratings. Therefore, the objective of a credit scoring system should be to target required deposits so they impact only the applicants most likely to default while imposing no financial burden on others.

A credit scoring table for one not-for-profit utility is illustrated in Table 3-5. This table is common in design to those for other not-for-profit utilities.

The utility depicted in Table 3-5 had been experiencing increasing bad debts for a few years. As a part of the efforts to address this negative situation, the utility's "current deposit" policy was customized as noted in the "Proposed Change" column in the table. The intent of this action, in conjunction with other efforts, was to put a stopgap on the increasing losses. At the same time, it was recognized that the utility's service rating most likely would decrease somewhat.

LESSON 8: CREDIT DATABASES

Modern utility-specific credit database matching services generate strong financial results.

Traditional credit scoring services are being enhanced and even replaced by modern multi-utility–shared credit database services. Utilities subscribing to these typically submit new service applicant, charge-off, and past due final data electronically in real time to a third-party company. Depending on the service provider, this information then

Table 3-5. Credit Scoring Table for a Not-for-Profit Utility

Credit Score	Number of Customers	% Customers by Score	Current Deposit	Deposits Collected	Proposed Change	Deposits to Collect
>860	6492	0.3231941	$0	$0	$0	$0
840–859	714	0.0355454	$0	$0	$0	$0
820–839	623	0.0310151	$0	$0	$0	$0
800–819	559	0.0278289	$0	$0	$0	$0
780–799	534	0.0265844	$0	$0	$0	$0
760–779	479	0.0238463	$100	$47,900	$100	$47,900
740–759	505	0.0251406	$100	$50,500	$100	$50,500
720–739	468	0.0232987	$100	$46,800	$100	$46,800
700–719	497	0.0247424	$100	$49,700	$100	$49,700
680–699	395	0.0196645	$100	$39,500	$150	$59,250
660–679	396	0.0197142	$100	$39,600	$150	$59,400
640–659	978	0.0486882	$100	$97,800	$150	$146,700
620–639	774	0.0385324	$200	$154,800	$200	$154,800
600–619	572	0.0284761	$200	$114,400	$200	$114,400
580–599	502	0.0249913	$200	$100,400	$200	$100,400
560–579	463	0.0230497	$200	$92,600	$200	$92,600
540–559	404	0.0201125	$200	$80,800	$250	$101,000
<540	4732	0.2355752	$200	$946,400	$250	$1,183,000
Total	20,087	1.0000		$1,861,200		$2,206,450

is made available to all participating utilities for the purposes of identifying new applicants who owe debts to former utilities as well as locating existing debtors.

Two companies providing database matching services are Equifax and ONLINE. For one year, Equifax reported that its service processed 31,443,844 customer service applications for telecommunications and utility providers. It matched 10.7% of these new service applicants to unpaid closed accounts in its database. Additionally, skiptracing unpaid closed accounts uncovered 10,328,647 debtors owing about $3,075,674,721.

ONLINE sells the Utility Exchange service in all 50 states. In its Southeastern region, debtors are identified when they apply for services at more than 400 utilities. The company combines the database information it compiles from subscribing utilities with information contained in its Rental Exchange database to maximize its credit assessment and collections performance.

Notes

1. Steve Maxwell, "How Big Is the 'Water Business'?" *Journal AWWA* (January 2005): 25–27.

4

Current Account Tasks

Once an account has been established with a local not-for-profit utility, customer and utility have a fairly routine relationship. Utilities provide their products and services and customers pay their bills. Most often, these bills are paid on a monthly basis. This chapter covers lessons learned regarding current account tasks including the hidden impact of inefficient billing decisions and considerations for addressing slow-paying accounts.

LESSON 9: BILLING FREQUENCY

Maximizing monthly billing frequency serves to minimize bad-debt expenses.

Bill presentment is a task required of all businesses that function as creditors. Two primary issues created by this process are (1) how to present customers with bills and (2) when to present those bills. Within the utility industry, a lot of attention has been paid to identifying the most cost-effective way to present bills—for good reason.

The costs associated with different methods for bill presentment vary greatly. Many utilities have completed estimations of bill-presentment costs. The results from one utility, Chugach Electric Association in Anchorage, Alaska, are representative of the results found throughout the industry. The cooperative estimated costs per bill were found to be $0.18 for e-billing, $1.61 for paper bills in the mail, and $10 for paper bills given in-person in the office.[1]

Obviously, a review of costs associated with various approaches to bill presentment is a worthwhile exercise for management. Likewise, a consideration of billing frequency is a worthwhile exercise for local officials/directors and management of not-for-profit utilities. A case example from one utility located in Texas illustrates this point.

Not-for-profit utilities use many different billing cycles (i.e., frequency), as shown in Table 4-1. The first utility noted has nine cycles. Cycles consist of several meter reading routes (e.g., five routes in one cycle). Whereas routes are a system for organizing readings, cycles are a way of organizing bill distribution. Individual readings are grouped into routes, and then routes are grouped into cycles to obtain economies of scale in the production costs resulting from printing and distributing bills.

A review of the billing and collections practices at the Texas utility in question uncovered a few key findings. The utility's write-offs were $967,072.30 (0.6% of revenues) for the preceding 2-year period. This amount exceeded the expected industry average of $490,783.93 by $476,288.37. A delayed billing-distribution process was identified as a major contributing factor. The billing process in place during the preceding 2-year period is given in Figure 4-1.

Table 4-1. Monthly Billing Frequency

Number of Customers	Billing Done by Utility or Outside Vendor	Number of Cycles
15,000	Utility	9
3,200	Utility	2
2,500	Utility	1
5,000	Vendor	8
2,500	Utility	1
15,000	Utility	18
15,000	Utility	3
7,000	Utility	6
12,000	Utility	4
19,000	Utility	20
25,000	Vendor	4

CURRENT ACCOUNT TASKS

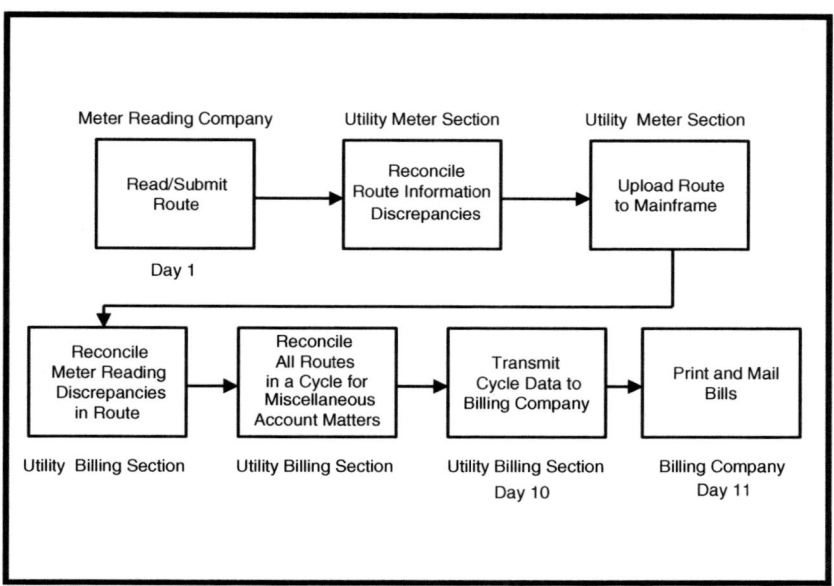

Figure 4-1. Current Billing Process for a Texas Utility

While executing its current billing process, the utility hired a contract meter-reading company that utilized SchlumbergerSema's RoadMAPS and RouteMAPS equipment and three personnel to read approximately 60,000 meters monthly. Contractual terms required that meters be read within 14 or fewer monthly business days, and that readings were to be delivered by 5:00 p.m. daily. The time to read all meters each month was 234 hours. Accordingly, it was estimated that the meters could be read in as little as 9.75 days per month (234 monthly hours/3 readers = 78 monthly hours per reader; 78 monthly hours per reader/8-hour day = 9.75 days per month).

The utility's meter section was responsible for coordinating all tasks with the contract meter reading company, as well as reconciling accounts and posting them to the utility's mainframe by route. Thereafter, the billing section further scrubbed the meter reading data with four personnel working for a combined total of approximately 40 hours per week. All routes were compiled in the billing section until a cycle was complete. Then the cycle was transmitted to the billing company for printing and distribution. Approximately 15,000 bills were distributed weekly. Based on the experiences of other not-for-profit utilities, the changes in Figure 4-2 were proposed to the utility's billing process.

The utility is reducing the time between when meters are read and when the corresponding accounts are billed. This change will decrease the utility's write-offs. By increasing the number of monthly billings from 4 to approximately 18, the utility would have decreased its bad debts by $84,377.77 during the preceding 2 years with a tighter timeline between read and disconnect dates. Less time means less usage and, correspondingly, lower bad debt. In addition, the utility will experience level cash flow rather than having payments arrive in four main lump sums each month. Last, workload in terms of customer interactions will be level as customers receive and pay their bills evenly over a monthly period.

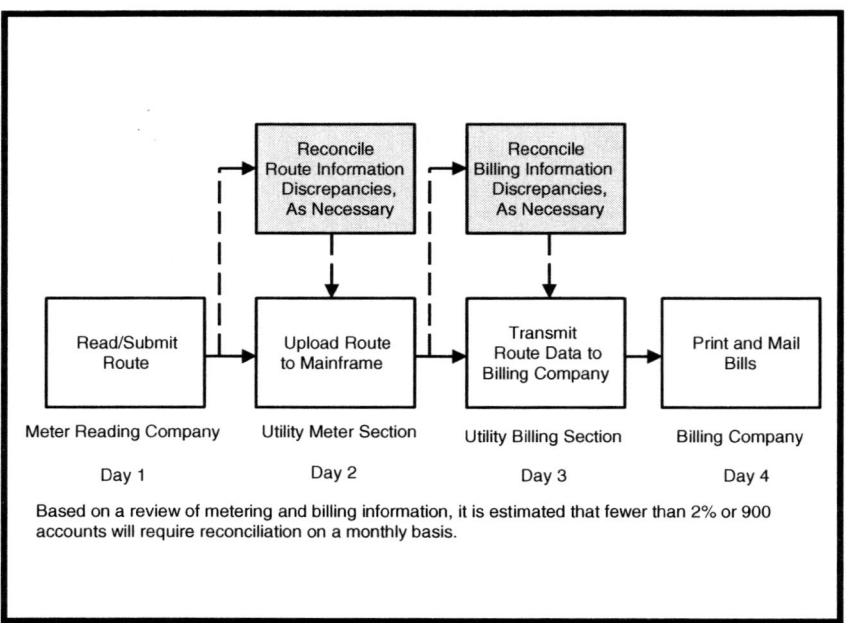

Figure 4-2. Recommended Billing Process for the Texas Utility

LESSON 10: MULTIPLE ACTIVE TIMELINES

Utilities can positively influence current accounts receivable and minimize bad debt losses by implementing variable timelines.

Most customer information systems enable utilities to vary the length of time they allow bills to go unpaid before taking action. This feature offers significant benefits because it makes possible the targeting of actions designed to encourage prompt payments and the limitation of bad debts to appropriate groups of customers. An example of how these systems are typically designed is provided in Table 4-2.

Rating systems segment customers into groups ranging from 2 to 5, for instance. Categorization is based on pay history and other factors. These systems facilitate prompt payments by customizing the information printed on bills and directing actions that are taken in response to late payments.

Another approach in practice that matches a billing process to a customer's bad-debt potential focuses on the timeliness of notifications and disconnections/shut-offs. A consulting project with one utility demonstrated these practices, as well as the need to curtail extended active timelines. Tables 4-3 and 4-4 provide an overview of the average billing timelines for the utility's customers as they were segmented into two categories: "friendly" and "termination."

Table 4-2. Typical Design of Timeline for Accounts Receivable

Rating	Criteria	Response to a Late Payment
1	On-time payment previous 24 months	No action
2	On-time payment previous 12 months	Mail notice
3	No history	Mail notice, phone contact
4	Late payments >0 and <3 previous 12 months	Disconnect/shut-off date printed on bill, phone contact
5	Late payments 3 or more previous 12 months	Disconnect/shut-off date printed on bill

Table 4-3. Average Billing Timeline—"Friendly" Category

Day	Activity
30	Meter read
40	Mail bill
56	Payment due
58	Late notice mailed
60	Meter read #2
70	Mail bill #2
86	Payment due #2
88	Late notice mailed #2
98	"Stated" shut-off date
108	"Actual" shut-off date

Table 4-4. Average Billing Timeline—"Termination" Category

Day	Activity
30	Meter read
40	Mail bill
56	Payment due
58	Late notice mailed
68	"Stated" shut-off date
78	"Actual" shut-off date

Again, this information demonstrates variable billing timelines at not-for-profit utilities. Unfortunately, these timelines were longer than those found at many comparable utilities from across the nation and abroad. Because the cost of an extended timeline is a lagging accounts receivable aging schedule as well as an increased bad-debt expense, the utility had to decrease these timelines to maximize the benefit it could receive from having them. The management of this utility opted to change its active timelines as indicated in Tables 4-5 and 4-6.

Table 4-5. New Average Billing Timeline—"Friendly" Category

Day	Activity
30	Meter read
33	Mail bill
48	Payment due
50	Late notice mailed with shut-off date
60	Meter read #2
65	"Actual" shut-off date

Table 4-6. New Average Billing Timeline—"Termination" Category

Day	Activity
30	Meter read
33	Mail bill with shut-off date listed
48	Payment due
53	"Actual" shut-off date

Multiple active timelines are made feasible by current technologies included with customer information systems. Utilities can influence their current accounts receivable positively and minimize their bad-debt losses by utilizing these technologies. For most utilities, using multiple timelines is simply a choice of incorporating the best business practices into their revenue management plans.

LESSON 11: BAD CHECKS

Bad-check policies with limits and appropriate fees minimize the number of checks utilities must process and the negative impact of these fees on service ratings.

Processing checks that are returned from financial institutions for reasons such as insufficient funds in a customer's account is a routine activity for not-for-profit utilities. In Table 4-7, a sampling of utilities in the Northwest illustrates the range of fees that utilities charge for processing bad checks, as well as policy limits placed on the number of bad checks a utility will accept. Once customers exceed the number of bad checks authorized within a 12-month period, they are required to pay bills in cash or with credit cards for the following 12-month period.

This study provided consideration for the impact a bad-check policy can have on the number of annual checks a utility must process and its service ratings. Of the utilities with no limits, 100% were given more than 0.02 bad checks per customer. In contrast, only 50% of utilities limiting the number to two in a 12-month period were given more than 0.02. The same results were obtained for utilities with a limit of three bad checks.

The overall opinion of customer service as it was related to the fee charged for bad checks is illustrated in Figure 4-3. Based on these results, not-for-profit utilities could effectively minimize the number of bad checks they have to process and avoid the corresponding costs incurred while doing so by establishing a policy with limits. Likewise, they could minimize the negative influence on service ratings that are associated with bad checks by limiting the fee charged for their processing. Identifying an appropriate fee ultimately requires balancing the costs incurred by the process with their effect on ratings.

Table 4-7. Range of Bad-Check Fees Charged by Northwest Utilities

Number of Customers	Fee ($)	Limit
29,000	10	2
6,000	50	2
17,000	20	2
55,000	20	2
3,000	0	3
50,000	25	3
3,000	10	3
110,000	50	2
3,000	6	None
17,000	7.50	None
18,000	25	2
17,000	20	3

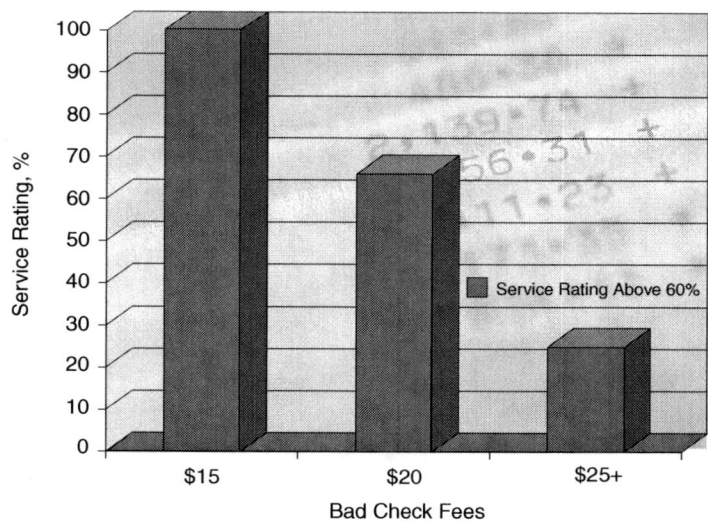

Figure 4-3. Opinion of Customer Service Related to Fee Charged for Bad Checks

LESSON 12: PAYMENT EXTENSIONS AND ARRANGEMENTS

Service ratings neither rise nor fall based on how many extensions or arrangements a utility provides, though authorizing more of them does lead to lower current accounts receivable.

Many not-for-profit utilities authorize customers to pay their bills after the established due date by granting them either an extension or an arrangement. Giving a customer five additional days within which to pay his or her bill prior to being disconnected is a typical *extension*. By contrast, allowing a customer to pay a balance due over several weeks or a few months is considered an *arrangement*.

An example of the number of not-for-profit utilities that offer extensions and/or arrangements is shown in Figure 4-4. This information was collected during a national collections web conference with more than 70 not-for-profit utilities.[2]

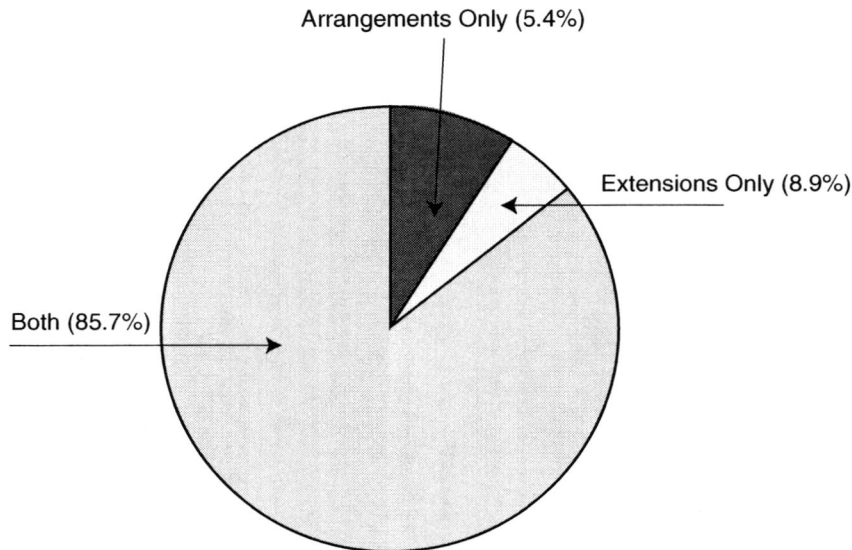

Figure 4-4. Not-for-Profit Utilities that Offer Extensions and/or Arrangements

Information provided by several not-for-profit utilities in the state of Georgia represents common polices established to manage the practice of payment extensions. None of the utilities attending this industry conference offered arrangements (see Table 4-8).

An effect from offering payment extensions and/or arrangements on service ratings was not present in the study results. Utilities that maintained more strict policies had similar service scores as utilities that had more liberal policies (see Table 4-9). In essence, making extensions and/or arrangements easier to obtain did not increase perceptions of service, although utilities with more liberal extension policies have been observed during consultation reviews to have lower current accounts receivable percentages than other utilities.

A policy is only as good as it is implemented. This statement is based on an old expression. It means that although several utilities may have the same policy regarding extensions or any other matter, their performance will vary depending on how a policy is implemented. A more thorough implementation or procedure generally leads to a truer

Table 4-8. Extension Policies of Utilities in Georgia

Number of Customers	Minimum Amount Owed ($)	Number of Extensions in 12-Month Period	Length of Extensions	Fee Charged for Extensions ($)
39,000	50	3	6 days	10
14,000	50	1 if broken	Next due date	0
90,000	100	3	7 days	0
17,500	100	1 if broken	Next due date	0
200,000	No minimum	4	7 days	0
33,000	30	2	Next due date	0
17,000	50	3	Next read date	0
125,000	30	4	7 days	0
40,000	50	2	Next due date	0

Table 4-9. Overall Opinion of Service by Number of Extensions and Arrangements per Year

Extension/ Arrangement per Customer	Utility Dispersion (%)	Customer Service Score (%)			
		<50	50–60	61–70	>70
<.15	50	20	20	40	20
>.15	50	20	20	60	0
	100				

measure of the impact a policy creates. The application for electric/water bill payment extension in Figure 4-5 is an example of one utility's procedure developed to carry out the policy set forth by its local officials. Note the extent of detail and process involved in qualifying for an extension at this utility.

```
APPLICATION FOR ELECTRIC/WATER BILL PAYMENT EXTENSION
                        CONFIDENTIAL
1.  NAME_____ ACCOUNT NO._____
2.  PRESENT ADDRESS_____
3.  TELEPHONE NUMBER_____
4.  REASON FOR EXTENSION REQUEST_____
    _____
5.  PLACE OF EMPLOYMENT_____
6.  OTHER SOURCES OF INCOME_____ HOW MUCH_____
7.  IF YOU ARE MARRIED, WHERE DOES YOUR SPOUSE WORK_____
8.  FAMILY SIZE_____
9.  HAVE YOU HAD YOUR ELECTRIC SERVICE DISCONNECTED FOR NON-PAYMENT
    WITHIN THE PAST TWELVE MONTHS_____ YES_____ NO

PROOF OF IDENTITY_____
I do hereby certify that the above information is true and accurate and is submitted for the
purpose of obtaining credit from City of Milford to extend the time for payment of amounts
due for prior electric service. I understand that if any of the above statements are found to be
untrue or false, I will be subject to immediate disconnection of services. I understand that
service can be disconnected for failure to comply with this agreement and for non-payment
for future electric service.


_____          _____
       Signature                              Date

_____
City of Milford Representative

                       PAYMENT SCHEDULE
                    Total amount of Bill    $_____
    Due Date                                Amount
_____       _____
_____       _____
```

Figure 4-5. Example of Application for Payment Extension

LESSON 13: LATE FEES

Nonexistent and inadequate late fees lead to lower average monthly current accounts receivable and increased losses.

Why do so many not-for-profit utilities charge a late fee when payments are not received by the due date? In short, the time value of money is the answer. The entire financial structure upon which a utility operates is based upon the simple expectation that customers will pay for the services they receive within a set time period established by each utility's local officials/directors. Typically, customers can expect to owe monies for services received on a monthly basis.

Monies not received from some customers within the prescribed time period negatively impact both the utility and, ultimately, the other customers. The combination of rates and fees at not-for-profit utilities are established to cover only the costs of providing its services. Accordingly, when budgeted monies are not received in a timely fashion from an expected source, they have to be collected through other means for the not-for-profit utility to remain solvent. In these cases, utilities are forced to operate with less than needed funds in the short run, and customers are forced to make up for this difference by paying more in rates and fees in the long run.

Two examples drawn from consultation projects with utilities demonstrate the negative impact the absence of late fees and the existence of insufficient late fees can have on not-for-profit utilities.

Utility Case 1: No Late Fee

A 5% or 10% immediate late fee and a 12% per annum continuing fee typically are found at most efficient utilities. These fees eliminate interest-free, short-term, and long-term loans in which bills are not paid in a timely fashion. It was identified that the not-for-profit utility in question did not charge either fee.

The absence of a late fee was driving at least two negative outcomes for the utility. First, "current" accounts receivable collections were lagging. The average current accounts receivable for the preceding 3 years ranged from 81% to 88%. At comparable utilities, these figures were found to range from 90% to 92%.

At best, the absence of an immediate late fee removed any incentive for customers to pay their bills by the due date. At worst, it actually encouraged them to "float" their bank accounts with interest-free funds from the utility during periods of tight cash flow (i.e., having more month than money).

In the most recent year, the operating revenues at the utility were approximately $88.9 million, and the current receivables were about 88%. If an immediate late fee had been in place, it was estimated that the current receivables would have been about 90%, leaving 10% or $8,890,000 to which a 10% late fee would have been applied. The fee would have generated about $889,000 in additional revenues.

The second negative outcome experienced by the utility involved the value of revenues collected. The absence of a per-annum fee decreased the present value of dollars collected from those customers who left the utility and returned in later years. A continuous interest penalty was needed to collect appropriate monies due and eliminate the current long-term interest-free loans the utility provided. For that reason, the utility added a 12% per-annum late fee.

Utility Case 2: Inadequate Late Fee

Another utility chose to apply to business accounts the same fixed $10 late fee it charged to residential customers. At this amount, many business customers were not paying their bills by the due date. After all, a $10 late fee did not seem that imposing when listed next to amounts due such as $3,000.

The monthly billings of the utility were about $3 million. Unfortunately, the utility's average current accounts receivable ran close to only 80%. The utility was constantly running into cash-flow problems because its business customers were paying approximately $600,000 late each month. And keep in mind that the utility was collecting only a $10 late fee per account. In sum, the utility collected about $1,000 in late fees each month.

After being advised of this situation, management of the utility decided it was time to change its late fee policy.

Notes

1. "The Real Cost of Billing," *Rural Electric Magazine,* (2004): 19.
2. Web conference by National Rural Electric Cooperative Association, April 14, 2004.

5

Delinquent Account Actions

Delinquent account actions come into play when a bill is not paid by its due date. The objective in these cases remains the same as that for accounts that do pay by the due date: the utility seeks to obtain payment in the least costly fashion while maintaining a high level of customer satisfaction. The key lessons in this chapter refer to the sequence of utility actions from disconnect/shut-off contacts to collecting unpaid accounts.

LESSON 14: AUTOMATIC PHONE-DIALER CONTACTS

The costs generated by notifying customers of a pending shut-off and the numbers of these actions are reduced by the use of phone-dialing systems.

Many utilities, as well as other businesses, use automatic phone-dialer systems—also known as predictive dialer systems—to contact customers. A utility's list of customers due for shut-off is uploaded to an information system that sends prerecorded messages by telephone, informing customers of a pending disruption to their service. Benefits to the use of these systems include "improved collection rates, considerable reduction of past due receivables, acceleration of cash flows, reduction in field collection activity, and a major reduction in overall collection costs."[1]

The following is typical of the messages communicated by automatic phone-dialer systems:

> Mr./Mrs. (Customer Name), please. This is your customer service representative with a friendly reminder from (Utility). Our records indicate that your payment of ($ Account Balance) is past due. This payment must be made by five o'clock p.m. tomorrow, (date), or (Utility) will have no alternative but to shut off your service. If shut off, the amount necessary to turn on your service is estimated to be ($ Amount).
>
> To avoid the inconvenience and additional expense associated with shutting off your service, you may pay with a credit card or check card by phone right now. Just press 1 on your telephone keypad.
>
> Thank you, (Customer Name), for your attention to this matter. Have a nice day.

Two examples of the impact these systems have had on utilities support the consideration of their use. One utility in Alabama that contracted for the use of an automatic-dialer system saw its number of shut-offs drop by almost 50%. In another case, a water and electric utility participating in the pilot study, Paris Board of Public Utilities, reported similar results. According to John Etheridge, office manager, after an automatic calling system was installed, "we saw a one-third drop in the number of customers on our cut-off list given to field representatives."

The cost to purchase or contract for the service of an automatic calling system varies based on the sophistication of the system. Erwin Utilities, another participant in the pilot study, reported spending about $3,500 to have the dialer software installed at the utility. "It costs us only about $150 a month to use the system," said Lee Brown, assistant general manager.

When compared to the cost of other forms of notification, dialer systems seem to be very competitive. Consider the cost of mailing late notices or shut-off notices. At one utility, the cost was determined to be $0.46 per mailed notice compared to only $0.08 per call. This meant that the utility was paying an additional $0.38 per contact before installing the dialing system.

DELINQUENT ACCOUNT ACTIONS

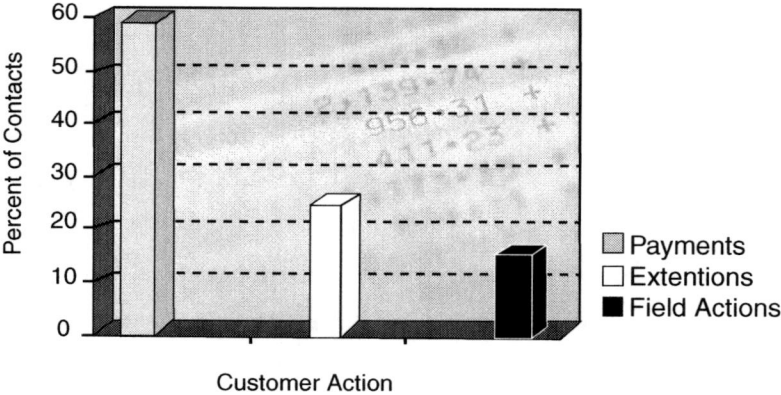

Figure 5-1. Response of One Utility to Phone-Dialer System

Last, measurements of the impact of dialing systems on collections performance and service have been found to be very positive. Typically, contact is made with someone at a customer's location about half of the time when calls are initiated by a system. Figure 5-1 illustrates the return one utility received from its investment in a phone-dialer system.

Comments provided by many industry veterans regarding dialing systems have led to the conclusion that the responses utilities receive from most customers is favorable. More often than not, customers report appreciating the system's contact as opposed to being annoyed by a telephone call. Negative reactions, albeit often loud, are the exception to the norm when it comes to customer reactions to automatic dialing systems.

LESSON 15: FIELD COLLECTIONS

The overwhelming majority of not-for-profit utilities collecting payments in the field report no incidents of theft or violence.

Whether utilities should maintain or adopt policies authorizing the acceptance of payments in the field is a highly debated issue within the industry. On one side of the debate, some argue that field collections should be based on consideration for customer service. As such, the matter is perceived as another opportunity that customers would have to pay their bills and avoid an unpleasant disconnection/shut-off.

The other side of the debate is based on security. The potential threat to utility personnel from their having to carry customer payments is viewed as an unnecessary risk. The solution that this group commonly advocates is, "Just make customers go to the office to pay."

A consideration of information regarding field collections provides some profound evidence upon which a utility can base its policy. Participants attending national, regional, and state association-organized collections courses; consulting projects; and an industry survey offer the same results about the matter. During association-oriented courses, participants routinely have been asked if anyone at their utilities has been the victim of an incident of theft or violence while collecting in the field. Only a few of the numerous participants recounted an incident ever having taken place at their utility. Many, many threats and occasional acts of violence were discussed that involved a disconnection/shut-off, but these occurrences were not associated with or instigated for a collector's holdings.

Consultation projects addressing utility collections have served to reinforce the anecdotal evidence provided during industry courses. Criminal behavior taking place in the field has been found to emanate from irate customers. Rarely has any suspicion of violence been motivated by theft.

Finally, survey information from utility representatives further confirms information available through industry courses and consultation projects. Figure 5-2 depicts survey data collected from representatives of 80 not-for-profit utilities. Among the utilities in this survey, 75% (60 utilities) reported that they "collect in the field and no one has been hurt/attacked for the money," 22.2% (18 utilities) "do not collect in the field," and only 2.8% (2 utilities) "collect in the field and have had an incident where someone was hurt/attacked for the money."[2] If only the 62 utilities that reported collecting in the field are considered, these data reveal that 96% (60 utilities) reported no incidents and 4%

DELINQUENT ACCOUNT ACTIONS

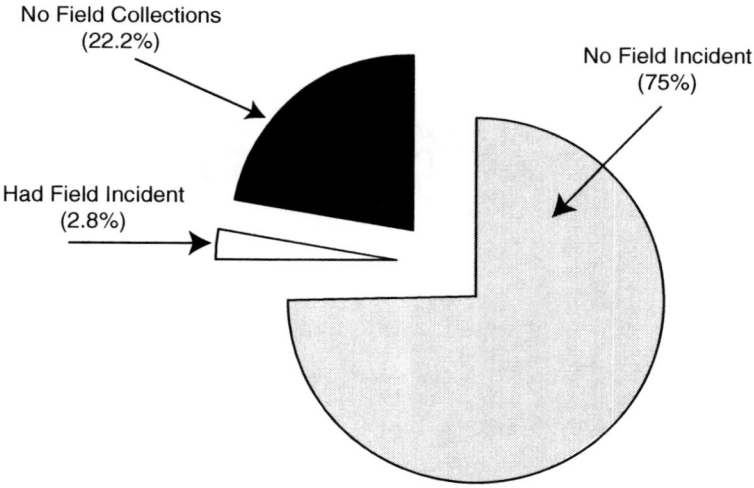

Figure 5-2. Field Collections Survey Data on Incidents of Theft or Violence

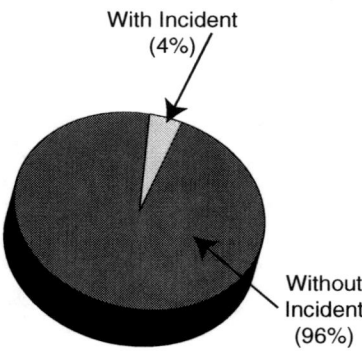

Figure 5-3. Incidents of Theft or Violence During Field Collections

(2 utilities) reported an act in which someone was hurt or attacked (see Figure 5-3).

Across the nation as a whole, incidents of violence or theft during field collections are not common. At utilities where incidents have occurred, anecdotal evidence suggests that the crime rate of the local areas was a major contributor to the probability of the episodes. Therefore, field collection policies might be best considered in light of an area's local crime activity.

LESSON 16: DISCONNECTION/SHUT-OFF

Improper information contained in shut-off notices can lead to legal and, thus, financial risk.

Just about the most volatile issue that water, electric, and natural gas utilities have to address concerns the shut-off of a customer's services for non-payment. Rarely are parties to the process happy about the reality of what is taking place. Local officials/directors and management of not-for-profit utilities face several delicate issues when crafting policies and operational procedures for carrying out shut-offs.

The means of shutting off services as well as associated fees are two prominent items that utilities have to address. In a poll of 61 utilities from across the United States, 68.4% of respondents reported that designated disconnect personnel carried out this task. The balance used combinations of meter readers, remote disconnect meters with electricity, and/or prepay meters.[3]

Some utilities deliver door hangers prior to a shut-off date, and others leave hangers on doors after they shut off service. The fees charged for delivering notices and those fees charged for carrying out a shut-off differ by utility. Table 5-1 gives an example of the fees charged by a group of Minnesota utilities.

As discussed in *The Utility Credit and Collection Manual*, "[E]ven though door hangers are a common and sometimes legally required notification procedure, utilities must use care when designing and handling these forms. Otherwise, they could run afoul with the legal system and incur costs resulting from issues such as promotion of a consumer's debts."[4]

Table 5-1. Door Hanger and Turn-On Fees Charged by a Group of Minnesota Utilities

Door Hanger Delivery		Turn-On		
Day or Night ($)	Number of People	Day ($)	Turn-On Night ($)	Number of People
25	1	75	75	2
30	2	30	100	2
20	2	50	150	2
25	1	50	150	2
25	1	25	60	2
10	1	25	60	1
35	2	120	150	2
30	2	40	40	2
20	1	60	160	2
10	2	20	75	2

What degree of exposure to litigation does the content of a utility's door hanger pose? This question is what local officials/directors and utility management must have answered by their local attorneys so they can make prudent decisions regarding this issue. Figure 5-4 illustrates ways of delivering information typically found on door hangers, as well as the likely degree of exposure each choice creates.

An example of a common door hanger is shown in Figure 5-5. It might present a debt-promotion problem for the utility if one of its notices is accidentally hung at an incorrect address or if the wind blows it off a door and into a neighbor's yard. Lawsuits have resulted from occurrences such as these in states including Georgia and Massachusetts. These expensive lessons were recounted by industry personnel during collections courses.

Perhaps the most expensive problem resulting from the ineffective use of door hangers involved the case of Civil Action 93-0004, *Francisca Osilek et al. v. Commonwealth Utilities Corporation (CUC) et al.* CUC is the municipal utility for the Mariana Islands, a US territory in the South

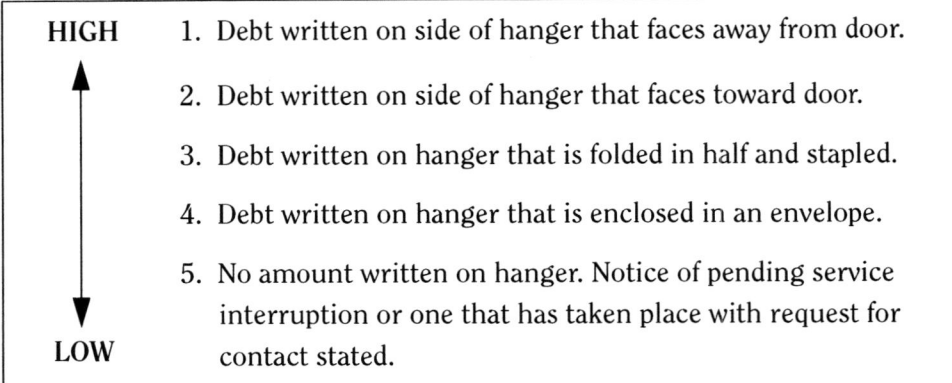

Figure 5-4. Legal Exposure From Information on Door Hangers

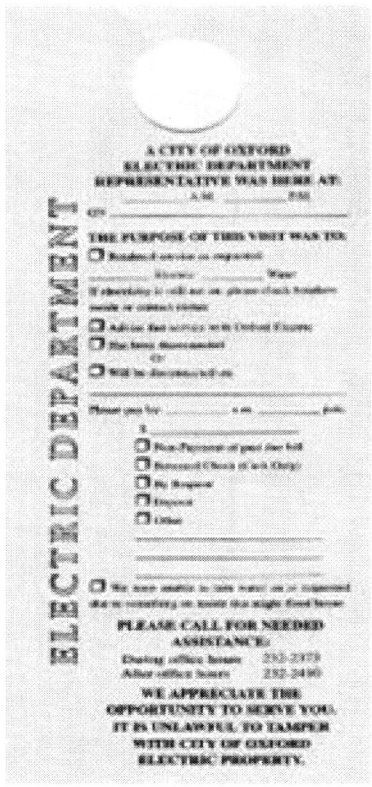

Figure 5-5. Example of Door Hanger

DELINQUENT ACCOUNT ACTIONS

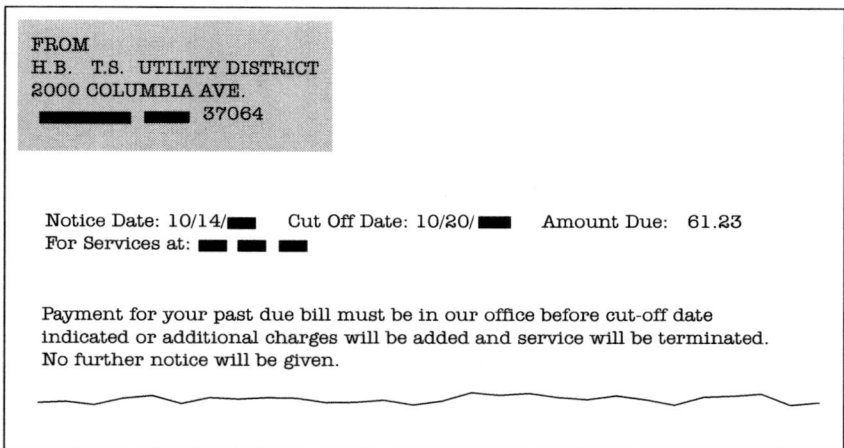

Figure 5-6. Example of Past-Due Bill

Pacific. As reported by the utility's former collections manager, a class action suit cost the utility more than $4 million when the courts found that the utility had embarrassed a class of its customers. The embarrassment resulted from the utility's use of a uniquely colored shut-off notice that most citizens came to recognize as the recipient of the hanger owing a debt.

The least risky shut-off notice might be one that isn't used at all. Some utilities opt to include a message with each bill stating that services will be discontinued if the bill is not paid by the due date. Others mail separate shut-off notices. The bill issued by H.B. & T.S. Utility District, a water utility, is in Figure 5-6 as an example of these notices.

LESSON 17: TECHNOLOGIES FOR DISCONNECTION/RECONNECTION

Technology has become the most cost-effective approach for performing many disconnections and reconnections.

From the "what I would like to do sometimes" desk: "Before it was purchased by Time Warner, Paragon Cable of New York City reported that collection of overdue bills improved dramatically when the company stopped punishing customers who owed money by cutting off service. Instead, deadbeat subscribers had their signals locked on C-SPAN.[5]

Which is more cost-effective—to have rows of people taking change in toll booths on turnpikes (i.e., toll roads) or for a state to use change-accepting machines with cameras to catch those who drive through without paying the toll? Would it cost an airline more to have rows and rows of ticket agents on hand to check in passengers or to provide automatic check-in stations? Finally, should most water utilities that also provide electricity continue to complete most of their disconnections by having employees drive around all day, or would it be more effective to deploy prepay and/or remote technologies?

As a result of advancements in electric metering technologies, water utilities that also provide electric service have new options available for doing traditional collections management actions. Although a few of the basic ideas behind these technologies have existed for decades, modern products available from vendors have made these approaches cost-effective for most utilities. First, consider the costs involved in having employees perform all of a utility's disconnections/shut-offs and reconnections/turn-ons. The cost information in Table 5-2 was drawn from the pilot study participants.

Keeping in mind that these costs reflect only wages and benefits and do not include items related to vehicle maintenance, etc., the average hourly rate for service personnel was $24.73 (i.e., $1,348,756/54,545). In addition to this traditional option of using service personnel, the three major technological options are: prepay meters, limiters, and remote devices.

Prepay Meters

Deployment of prepay electric meters by utilities has been undertaken for a variety of reasons. The Orangeburg Department of Public Utilities in South Carolina began offering prepay meters to its customers after the utility increased its security deposit policy. Customers could choose this

Table 5-2. Cost of Disconnection and Reconnection Performed by Employees

Utility	A Year of Wages and Benefits	Hours During Year Working
1	192,000	7,680
2	202,187	6,250
3	17,442	850
4	47,112	2,400
5	56,000	1,575
6	6,980	240
7	374,085	15,300
8	81,450	5,000
9	182,500	8,250
10	189,000	7,000
	$1,348,756	54,545

option to avoid paying the upfront deposit. At Florida Power & Light, as well as Carolina Power & Light, customers could elect to have a prepay meter installed to avoid a pending disconnection. Roughly 95% of customers chose the prepay meters. And, after being given the option to go back to their old meters or to stay with the prepay meters, 75% chose the prepay meters.[6] Thus, customer satisfaction was reportedly positive when prepay electric meters were involved.

A few years ago, Louisville Gas and Electric in Kentucky conducted a pilot program with 500 prepay meters. When asked to rate their experience with the meters, 85% of customers reported extremely positive experiences.[7] Another case involving Salt River Project Utility in Arizona produced similar results. According to a report in *Transmission & Distribution World*:

> Surveys of customers in the program revealed that about 92 percent were satisfied with it and many were very satisfied, giving PAYGo the highest

level of satisfaction of any program so far offered by the utility. About 75 percent of these customers indicated they had a higher opinion of SRP after participating in the program. This result was especially notable, since many of the participants entered the program with a negative view of the utility, after having experienced late-payment fees, disconnections and increased deposits. [8]

Unfortunately, electric prepay meters remain more expensive than other technical options such as limiters and remote disconnect devices. Primarily for this reason, prepay systems have not been widely deployed in the United States. But in countries such as Great Britain, more than 4 million customers use electric prepay systems and approximately 2,500, or 25%, of Woodstock Hydro Services' residential customers in Ontario, Canada, have chosen electric prepay systems over the traditional billing process.[9,10]

Limiters

Limiters are devices that curtail the amount of electricity that will flow through a meter. Over the years, limiters have been used for a variety of reasons including the management of critical-care customers. For instance, Rochester Public Utilities in Minnesota reportedly uses limiters to minimize the financial risk of continuing to provide electricity to customers with delinquent accounts who have specific medical conditions.

Remote Devices

Most recently, remote technologies for performing disconnections and reconnections have become available. Companies such as BLP Components offer a variety of means for performing these services remotely from a utility's offices. For instance, BLP's Powerpulse consists of a collar placed between any type of meter and its meter base. The company's network software enables a utility to turn off meters using simple paging signal technology. The collars cost about $200 to $240 each, depending on the number ordered.

Ultimately, the decision to use employees or to use technology to perform disconnection and reconnection services is based mostly on cost. Considering the information made available from the pilot study, a quick calculation will reveal the payback period for any investment in remote technologies. Using BLP's x-Pulse Payback Calculator (www.blpx-pulse.com), it was determined that the payback period would be approximately 2 years.

When considering the optimal number of remote disconnect devices a utility should purchase, a modified form of breakeven analysis is a useful exercise. Mike Sells, administrative director at Orangeburg, South Carolina Department of Public Utilities, used this process in determining the number of units to purchase (see Table 5-3).

To project the approximate number of remote devices that Orangeburg could use, disconnect data from the most recent 5-year period were analyzed. According to this information, the breakeven point between the deployment of remote technology and the assignment of personnel to perform disconnects is at about 1,000 units (number of customers = 530 + 470). Here the utility would spend $250,000 ($132,500 + $117,500) for remote devices and achieve $253,218 in personnel savings ($164,290 + $88,928).

LESSON 18: DECONSTRUCTING BAD DEBTS

Calculating the average daily bad debt is the first step toward deconstructing aggregate losses.

Managers of not-for-profit water, wastewater, electric, and natural gas utilities are periodically asked to explain the forces contributing to a utility's bad debts. The case example provided in the ensuing discussion includes a methodology for deconstructing bad debts for the purpose of revealing these forces. This case example is drawn from an actual consultation project involving an electric cooperative.

Table 5-3. Examination of Applying Disconnect Device for Disconnects Over a 5-Year Period

Number of Disconnects per Customer	Number of Customers	Total Number of Disconnects	% of Total Disconnects	Total Cost at $250 per Remote Unit	Disconnect Fees at $30 per Incident	Personnel Savings at $14 per Incident
1	5,503	5,503	10.6	$1,375,750	$165,090	$77,042
2	2,470	4,940	9.5	$617,500	$148,200	$69,160
3	1,521	4,563	8.8	$380,250	$136,890	$63,882
4	987	3,948	7.6	$246,750	$118,440	$55,272
5	703	3,515	6.8	$175,750	$105,450	$49,210
6–7	858	5,522	10.6	$214,500	$165,660	$77,308
8–10	680	5,993	11.5	$170,000	$179,790	$83,902
11–14	470	6,352	12.2	$117,500	$190,560	$88,928
15–84	530	11,735	22.5	$132,500	$352,050	$164,290
Totals	13,722	52,071	100			

Background

The objective of this project was to perform a comprehensive review of "Hometown Electric Membership Cooperative's" (HEMC's) credit and collections activities and to offer recommendations for optimal performance. Factors contributing to the cooperative's growing bad debts were given particular attention.

The cooperative's revenues grew by 8% between year 1 and year 2 and 13% between year 2 and year 3 during a 3-year period. Unfortunately, the utility's growth in bad debts outpaced its growth in revenues. Bad debts grew by 14% during each of those years. Table 5-4 shows these changes.

The growth rate of delinquent monthly average accounts receivable balances negatively supported the utility's increasing bad debts during the past couple of years. In fact, this rate of growth doubled from 10% between years 1 and 2 to 20% between years 2 and 3 (see Table 5-5).

HEMC's collections management was moving in a direction toward ever-increasing bad debts that were fed by an expanding accounts

Table 5-4. HEMC's Revenues and Bad Debts, Years 1–3

Year	Revenue ($)	% Change	Bad Debt (%)	% Change
1	209,043,401		37	
2	225,015,636	8	42	14
3	254,698,536	13	48	14

Table 5-5. HEMC's Growth Rate of Delinquent Accounts Receivable, Years 1–3

Year	Current $	12-Month Average ($)	% Change	30+ Days	12-Month Average ($)	% Change
1	149,036,229	12,419,685	12	47,789,841	3,982,485	
2	166,239,333	13,853,277		52,686,156	4,390,513	10
3	195,483,993	16,290,332	18	63,252,730	5,271,061	20

BALANCING COLLECTIONS PERFORMANCE AND SERVICE RATINGS

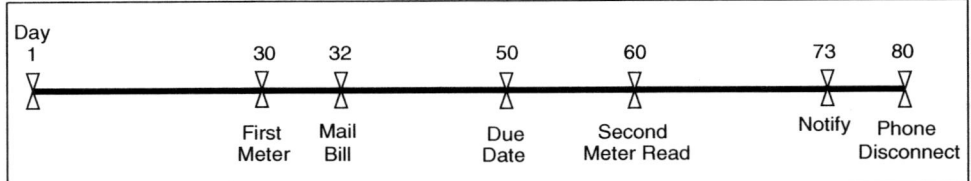

Figure 5-7. HEMC's Active Account Management

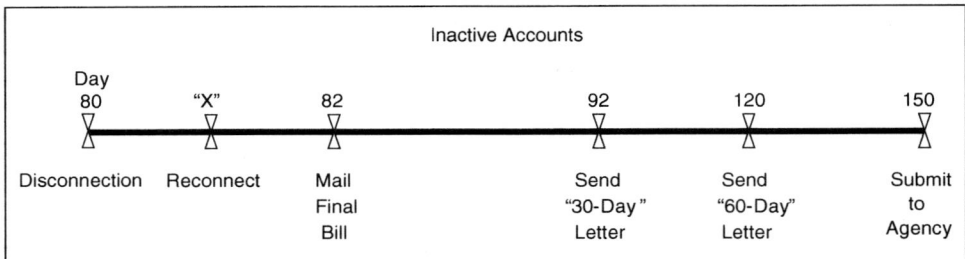

Figure 5-8. HEMC's Inactive Account Management

receivable schedule. Accordingly, the cooperative's board and management promptly began to implement the recommendations described below.

Observations

The timelines in Figures 5-7 and 5-8 illustrate HEMC's active and inactive account management. Although the utility's policies allowed for disconnections on day 75, in practice this action was found to occur on average on day 80. This is a common observation.

Recommendations

These timelines, in conjunction with the deployment of certain technologies and the execution of various business practices (e.g., payment agreements), resulted in $1,229,753 in bad debts, or .48% of revenues for the most recent year. Residential accounts generated 79.02% of these losses, and only 20.98% came from commercial accounts.

DELINQUENT ACCOUNT ACTIONS

A review of the myriad of financial and operational reports associated with HEMC's account management activities, as well as consideration of lessons learned from reviewing collections practices across the utility industry, led to the following conclusion regarding the cooperative's bad debts. The four primary factors that combined to create the utility's losses were:

1. .12% active timeline
2. .07% inactive timeline
3. .19% deposit threshold
4. .1% miscellaneous

These factors and the degree of influence are illustrated in Figure 5-9.

.12% Active Timeline

HEMC wrote off $1,229,753 in the most recent year. Dividing this figure by the average number of days an account remained active prior to disconnection (i.e., 80) provided the average daily bad debt. This amount was $15,371.91.

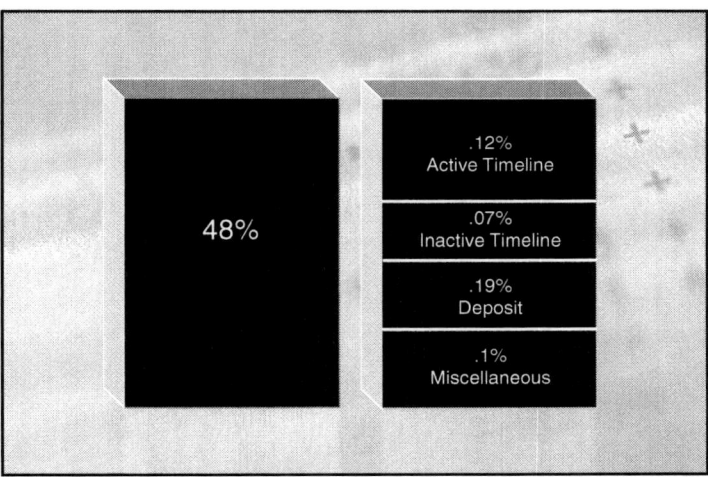

Figure 5-9. Deconstruction of HEMC's Bad Debts

The first recommendation was that HEMC change its disconnection policy. Rather than disconnecting all accounts 45 days after a bill was due, it was offered that the cooperative maintain two disconnection timelines. High-risk accounts indicated by either a poor credit rating or three late notices within the previous 12-month period would be subject to disconnection within 30 days from their due date. These disconnections would take place during the meter reader's normal routes.

Low-risk accounts or all others would be subject to disconnection 60 days from their due date. Again, disconnections would take place during the meter reader's normal routes.

Previously, HEMC attempted to use the same personnel to read the cooperative's meters and then, as time permitted, conduct disconnections. This practice seemed to produce a negative consequence.

After completing their routes, meter readers were required to disconnect meters located at a variety of places across each district's service area. Because of the geographical distances, delays in carrying out disconnections were prevalent. This was serving to increase the utility's ultimate losses.

Moving forward the time when high-risk customers were subject to disconnection while moving back the same timeline for low-risk customers was suggested as a way to balance HEMC's customer service and collections requirements. Customers who were timely payors would be affected positively by this change, and the impact upon collections activities would be only the addition of 10 days (i.e., going from 80 to 90 days prior to a disconnection).

Conversely, the impact on high-risk customers who traditionally had the dominant impact on bad debts was seen as financially beneficial to HEMC. It was suggested that their disconnection timeline should be moved forward from 80 to 60 days.

The influence on the utility's bad debts of disconnecting high-risk accounts at 80 days rather than the suggested 60 days was $307,438.25,

DELINQUENT ACCOUNT ACTIONS

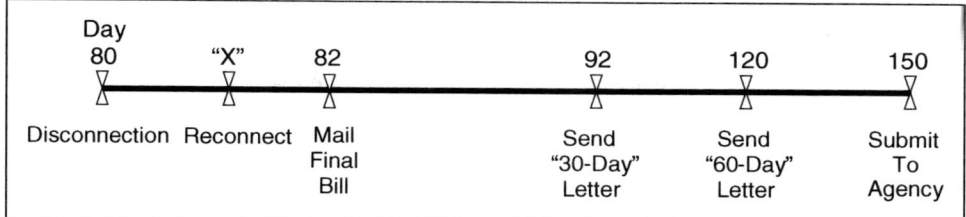

Figure 5-10. Inactive Accounts Collection

or .12% of revenues. This amount was determined by multiplying the 20 extra days that accounts remained active (i.e., 80–60) by the average daily bad debt amount ($15,371.91) determined previously.

.07% Inactive Timeline

HEMC undertook a series of communications with consumers in default between the point of disconnection and collections. These communications included phone calls, a final bill, a 30-day letter, and a 60-day letter. For purposes of discussion, the inactive timeline is reproduced in Figure 5-10.

The extended time period between "Disconnection" and "Submit to Agency" dates was having a significant, negative impact on the utility's bad debts. The key obstacle to collection of debts is *time*. Monies become less likely to be collected over time because debtors move, commit fraud, and so on. Therefore, minimizing time maximizes collections.

HEMC allowed for 70 days between disconnection and the commencement of collections. In the most recent year, the utility submitted $1,135,135 to its agency for collections and received $343,821 in monies collected. This equates to a 30% success rate. It was suggested that HEMC condense its timeline for inactive accounts collection as illustrated in Figure 5-11.

The suggestion to submit accounts to collections 60 days earlier than was the practice was projected to influence bad debts in a profound way. Again, in the most recent year, the cooperative received $343,821, or a 30% rate of return, while maintaining a 70-day disconnect-to-submission

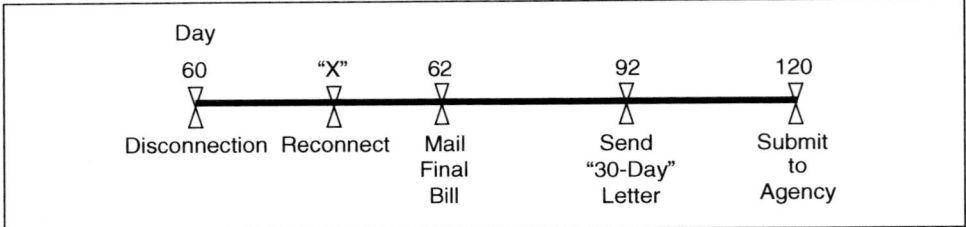

Figure 5-11. Proposed Inactive Accounts Collection

time period. It was important to emphasize that the success rate would be directly contingent upon this timeline. Reducing the time to less than 70 days would reduce the utility's bad debts correspondingly.

Considering the cooperative's investments in credit assessment technologies and its extensive collection of information during the applications process, if HEMC had maintained a 60-day period, it would have achieved approximately a 45% success rate in collections. This figure is unusually high regarding agency returns based on experience with utilities and collection industry data. Most utilities will not see results this high.

The difference between what could have been HEMC's results (i.e., $510,810 = 45% success rate) and what it received (i.e., $343,821 = 30%) was $166,989. This dollar amount constituted .07% of the utility's bad debts.

.19% Deposit Threshold

Holding constant the issues of active timeline, inactive timeline, and miscellaneous contributors to bad debts, the remaining .19% was recognized as having been created by the difference between accumulated debts and available security deposits. This percentage contributed $486,777 to bad debt.

Implementation of the recommendations regarding active timeline and inactive timeline reductions was projected to simultaneously provide closer balance to the amount of security on-hand relative to defaults. Reducing the amount of outstanding balances at the time of disconnection and increasing the amount of funds collected would minimize the need to collect and maintain security deposits.

In the most recent year, the utility had $1,229,753 in bad debts. Implementing the active and inactive timeline adjustments would have reduced this amount by $474,427 to $755,326. HEMC would have had to collect this amount in additional security deposits to cover the shortfall.

It was reported that 5,002 accounts contributed to bad debts. Therefore, an estimate of the deposit shortfall per account was $1,229,753/5002 = $245.85 per account. Taking the active and inactive changes into account, this amount would have been $755,326/5002 = $151 per account. Any combination of increased deposit amounts or adjustments to the credit scoring formula that generated these funds would have resolved this shortfall.

.1% Miscellaneous

In the ideal world, usage would be calculated, billed for, and paid in full. Unfortunately, historical data and experience prove that this does not happen. Therefore, a .1% cushion normally is advanced as an unavoidable cost of doing business. In some cases, it may not be completely unavoidable, but the cost to fully collect these monies is prohibitive.

LESSON 19: IN-HOUSE VERSUS OUTSOURCING COLLECTIONS

Information available primarily via the Internet makes locating debtors a relatively low-cost task for most utility collectors.

A portion of the collections flowchart from chapter 1 is reproduced as Figure 5-12. When bills go unpaid, most utilities chose from a distinct set of options to pursue collections. Across the industry, figures vary as to what percentage of utilities chose which pathways. In one survey, 79.5% of respondents stated that their not-for-profit utility's in-house personnel initially attempt to locate and collect from debtors.[3]

BALANCING COLLECTIONS PERFORMANCE AND SERVICE RATINGS

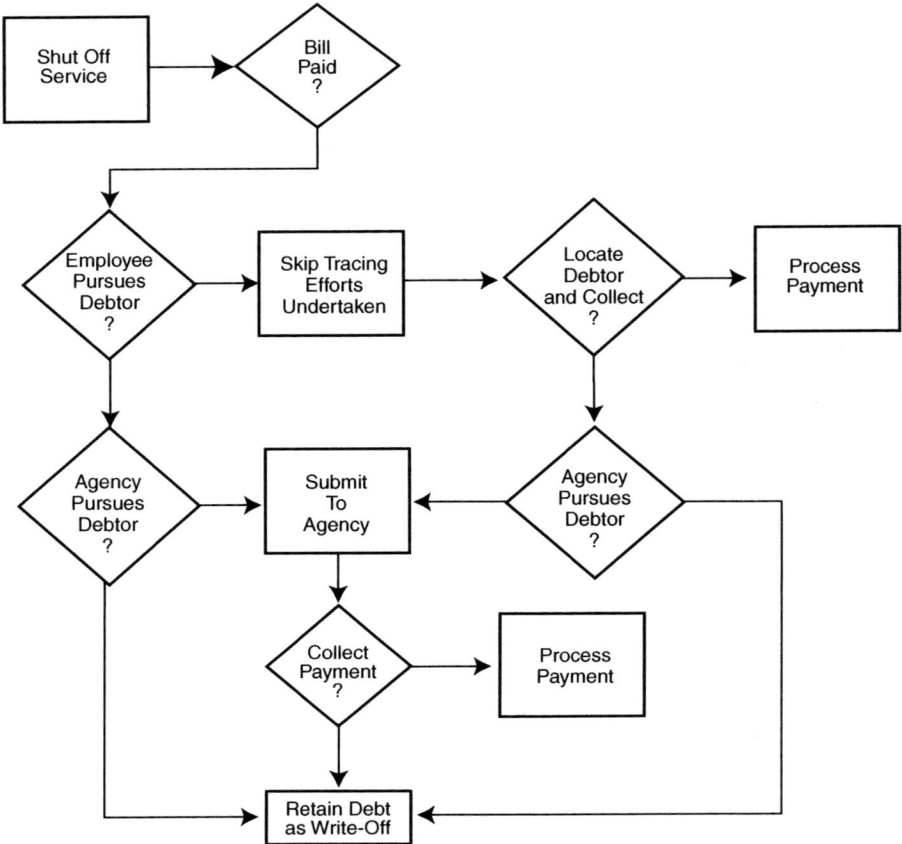

Figure 5-12. Portion of Collections Flowchart

The results of in-house collections efforts are found to be effective at some utilities and less so at others. Data collected during the pilot study reveals the dollars pursued by the group's employee collections personnel as well as their effectiveness. In sum, the employees were assigned $3,161,499 in a recent year and were able to collect $693,659, achieving a 21.94% collection ratio (see Figure 5-13).

Normally, the decision by local officials/directors and management on how to pursue collections rests upon the same considerations involved with most of the other matters covered in this report: financial performance and service ratings. Individual utilities must perform a cost-analysis of their own operations to determine if the investment in

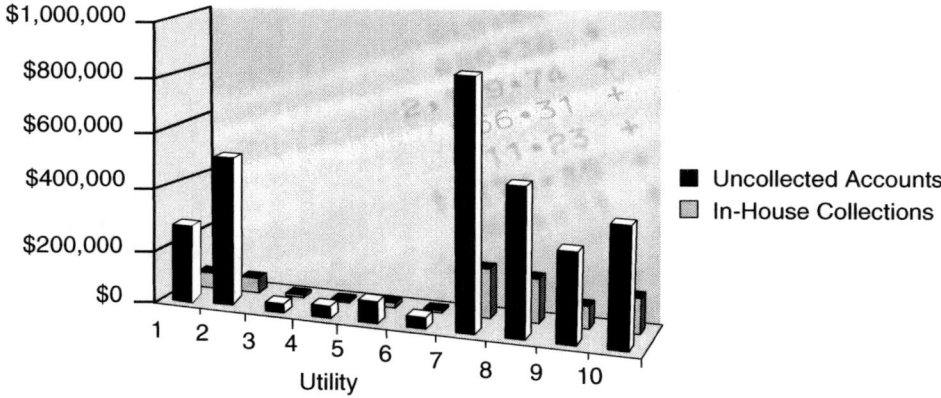

Figure 5-13. In-House Collections in Pilot Study

personnel and resources necessary to pursue collections makes this a viable option. Regarding service ratings, as detailed in chapter 2, lessons 1 and 3, utilities that maintained a higher priority on collections in the pilot study did not have lower satisfaction scores than other utilities. Therefore, in-house efforts seem to be supported as far as customer satisfaction is concerned.

In practice, collection of debt is a two-part process: locate and procure. Before monies can be obtained, a debtor must be located. Skip tracing is the common name for the process of locating individuals. Location information available primarily via the Internet makes finding debtors a relatively low-cost task for most in-house collections employees.

Companies that offer location information continue to enter the marketplace every year. And existing companies revise their mix of services and prices quite often. For the latest providers, options, and prices available, collections personnel should consult Internet search engines. A few popular resources accessed for locating debtors include:

Accurint.com
Quickinfo.net
Search.com

Lexis/Nexis: Bankruptcy (BANKO)
Experian.com (Electronic Directory Assistance)
National Change of Address (from USPS)
Social Security Death Index (SSDI)
Directory.net

Many vendors include free trial runs through their Web sites. Some companies price their services per search (e.g., Accurint), and others offer flat-rate prices for fixed periods of time (e.g., QuickInfo).

Irrespective of whether the physical address or phone number of a debtor is found, some utilities opt to place debt information directly with credit bureaus. Thereafter, to improve their credit worth, many debtors contact the utilities to pay the monies owed. Utilities interested in starting this process need only contact one or any combination of the three major credit bureaus—Experian, TransUnion, and/or Equifax. The bureaus' representatives will indicate whether any fee would be involved in the process.

Once debtors are found, step 2, procurement of funds, takes place. Three primary options used by utilities that are effective at collections are letters, court actions, and tax liens. Letters informing "lost" debtors that a utility knows where they have moved, as well as its intention to take collection actions, prove effective with a healthy percentage of debtors. Suing debtors in small claims court has the potential to draw in another considerable portion of monies owed. Last, states such as South Carolina allow utilities to place liens on income tax refund checks for unpaid bills.

LESSON 20: COLLECTIONS AGENCY REQUEST FOR PROPOSALS

The first major reason many utilities get inadequate results from collections agencies is the absence of a formal selection process.

Contracting with an agency for collecting delinquent final bill debts is a common utility practice. By far, the majority of utilities represented at state, regional, and national association collections courses report using the services of collections agencies. Regrettably, many of these utilities experience far from satisfactory performance from their agencies.

Two primary reasons have been identified as to why so many utilities receive unsatisfactory and even poor results from their collections agencies. The first reason is included in this lesson, and the second reason is covered in the next one. In many cases, the first reason that utilities get inadequate results is the absence of a formal selection process. A utility is a major organization when compared to a local dentist's office, for instance. Small collections agencies are appropriate for these businesses because in most cases they simply do not have the personnel, technologies, or financial resources to serve the collections needs of a utility.

To obtain appropriate collections services, a utility should release a request for proposals (RFP). Here we explore a case example of the process that one utility followed for developing and releasing an RFP, as well as the process it took to evaluate prospective agencies.

Utility Case Example: Collection Agency Search

The senior management of a US not-for-profit utility initiated a consultation project designed, in part, to select a well-qualified agency to perform collections services. The management conducted a review of the utility's in-house collections processes, its people and technology, and the priorities of its management concerning financial and service performance. Based on this utility's situation, a specific RFP was developed. A copy of the RFP is included as appendix B.

The agencies were given about a month between the release date of the RFP and the submission date for proposals. Thirteen collections agencies from across the country received copies of the RFP, and eight submitted responses. A general overview of the similarities and differences of the respondents is presented in Table 5-6.

Table 5-6. Similarities and Differences Among Collection Agency Respondents

Similarities	Differences
All had over 10 years of experience	Two agencies listed non-profit utilities as existing clients
Sample reports were very comparable	Most target investor-owned utilities
Six of eight report debts to the three main credit agencies 30 days after receiving accounts	Better Business Bureau review indicated problems with two agencies
All send first letter within 3 days of receiving accounts and begin calls between 1 and 7 days	Fees and terms varied (e.g., 18 30% commissions)
Monies are recovered up to 6 months; thereafter, little results occur	One agency carried service-related citations

A detailed review of the proposals served to separate the best candidates for the utility's specific situation from the rest of the candidates. The several criteria examined included:
- experience working with utilities,
- Better Business Bureau review,
- proposed fees,
- proposed terms and conditions regarding length of contract,
- resources available to perform the services proposed, and
- presentation quality of each proposal.

A table of data was constructed (see Table 5-7) to see how each agency responded to certain key items.

Armed with data collected from the proposals, the utility's management was in the position to make a logical, informed choice based on the needs of its utility and the capabilities of the respondents. In the end, three agencies were identified for consideration, and the remaining five were found to be inappropriate. Once the appropriate agencies were separated from the total pool of candidates, the second step

Table 5-7. Key Items for Analysis of Collection Agencies

	1	2	3	4	5	6	7	8
Location	MA	TX	TX	MN	TX	OK, AL, & MS	PN	IL
Year founded	1967	1968	1980	1981	not stated	1980	1955	not stated
Utility references	2 muni, 3 iou	none	2 muni, 1 iou	1 coop, 2 iou	3 coops	1 muni	7 muni, 6 iou	3 iou
Better Business Bureau check	satisfactory record	unsatisfactory record	satisfactory record	no record	satisfactory record	satisfactory record	unsatisfactory record	satisfactory record
Citations outstanding	none	none	none	none	none	none	none	2 citations
Number of collectors	20	40	45–50	>100	22	191	130	955
Average number of accounts per collector	2,500	250–300	not specified	300–500	3,500	not specified	800–1200	750
Organizationl chart present	yes	yes	yes	yes	none	none	yes	yes
Commissions	25% = acct<1 yr; 50% = acct>1 yr	18%; second lacement 35%	21.90%	24%	28%	30% no litigation; 40% with litigation	21%	18%
Extra fees	legal costs	legal costs	legal not specified	legal costs	legal not specified	none	legal not specified	litigation on case-by-case
Sample reports	yes	yes	yes	yes	yes	yes	yes	yes

(continued next page)

Table 5-7. Key Items for Analysis of Collection Agencies (continued)

	1	2	3	4	5	6	7	8
Contact process	1st letter immediately call in 7 days 2nd letter after 30 days 2nd call after 30 days report to Experian on day 30	unspecified letters and calls report to credit agencies	letter and call immediately 2nd letter after 20 days report to 3 credit agencies	Varies by account $ calls daily to every 2 days 5 letter series report to 3 credit agencies	4 letter series day 8 calls begin calls continue over time	3 to 5 letters calls begin within 5 days report to credit agencies	2 letters series of calls report to 3 credit agencies	calls every 7 days 3 letter series
Average collection time	returns account after 9–12 months	122 days	within 6 months	67 days	none specified	maximum 9–12 months	none specified	within 120 days
Term of service	as per utility	1 year	as per utility	6 months	18 months	as per utility	as per utility	as per utility
Signature	yes	yes	yes	yes	yes	yes	yes	yes
Miscellaneous	none	none	30% bilingual	bilingual collectors	none	1998-#13 agency by rev.	none	returns accts 6–12 mo

in the process of contracting for effective collections services—crafting terms—could be undertaken.

LESSON 21: COLLECTIONS CONTRACT TERMS

The second major reason many utilities experience inadequate results from collections agencies is that contractual terms do not support positive results.

Among the many terms contained within a collections agency's contract, a few key items carry the bulk of the weight when it comes to financial performance. The terms impacting utility revenues that have to be considered carefully are commissions, fees, and duration. A sample of various commissions established between utilities from across the United States and their collections agencies is provided for review in Table 5-8. There is simply no justification for commissions over 30%.

Various commission percentages are paid by the utilities listed in Table 5-8. In addition, payment conditions are attached to several of the commissions. If a debtor pays within the first 30 days of placement, for instance, some of the utilities are not obligated to pay a commission to their agencies. The rationale behind this is that the debtors were going to pay and just happened to have been placed with the agency prior to that date. But it is worth considering that this term might be counter-productive in providing an incentive for collectors to delay their actions until the zero commission time period lapses.

Perhaps the most controversial issue surrounding a commission is not its amount, but rather, which party is directed to pay it. One utility listed in Table 5-8, as well as two from the pilot study and others across the country, maintain policies whereby debtors pay all costs associated with collections. Accordingly, each time an agency is assigned a debt, $100 for example, the entire amount is returned to the utility and the debtor pays the commission to the collector.

Table 5-8. Collections Agencies' Commissions

Number of Customers	Commission %	Number of Customers	Commission %
7,500	33.3	16,000	30
8,200	33.3	16,000	0 within 30 days, then 40
3,800	50	2,500	20 within 30 days, then 40
16,000	0 with letter, then 30	31,000	0 within 30 days, then 40
11,000	30	3,000 (self reports debts)	0 within 30 days, then 40
42,000	30 or 0 at utility >6 mo	15,000	40
20,000	40	10,000	40
30,000	40	3,300	40
10,000	40	15,000	40
22,000	40	43,000	40 debtor pays
11,000	50	30,000	40
5,000	0 with three letters, then 33.3	10,000	0 within 30 days, then 40
5,300	50	85,000	15 or 0 at utility
28,000	25 within 30 days, then 40	17,000	40 or 0 at utility

A second term for consideration is fees. The commissions listed in the table do not include legal fees for suits brought by the agencies on behalf of the utilities. A higher commission percentage or fee is common in these situations. If a utility has a flat 30% commission for non-litigated collections, it also might have a 40% commission for situations involving litigation, but other types of fees should be viewed skeptically.

A third term involves the duration of time an agency is authorized to pursue a debt. Debtors pay approximately 75% or more of monies gained through the collections process within the first 6 months of placement with an agency. After this, most debtors pay only to clear their credit reports in relation to some purchase they are attempting to make. In short, they are not paying as a result of any efforts made by a collection agency. Some utilities find it prudent to place accounts with agencies for 6 months to 1 year and then require all uncollected accounts to be returned to the utility. Accounts are returned so utilities do not pay commissions needlessly when debtors attempt to clear their credit reports or relocate back to the utility's service area.

Utility Case Example...Continued

Once the utility discussed in lesson 20 made its agency selection, the next step was to establish an agreement for services. A benefits projection was constructed to determine the outcome of three key matters. The matters under review consisted of the following:

- *Establish "straight" contract:* Utility contracts for services with no time limits and pays commission as proposed by respondent.
- *Set time limit:* Recall accounts from selected agency after a set period of time.
- *Debtor pays commission:* Agency collects funds owed utility, as well as collection costs from debtor.

Projections were prepared, taking into consideration that the utility had approximately $500,000 in yearly debts available for placement, the selected collections agency proposed a 25% commission, and it was anticipated that the agency would actually collect 35% of the monies outstanding. Based on these figures, the outcomes shown in Figures 5-14, 5-15, and 5-16 were estimated.

The example utility had gone through the formal process of selecting a suitable collections agency (see lesson 20), investigated the impact of alternative terms, and was fully prepared to enter into an agreement with an agency. The rewards to the management and staff of this utility for their thoroughness are a high collections rate of success (i.e., 35%), maximum net returns (i.e., debtors paying commissions), and fewer-than-normal customer complaints regarding collections (i.e., professional agency selected). The utility, agency, and customers all benefit from the process.

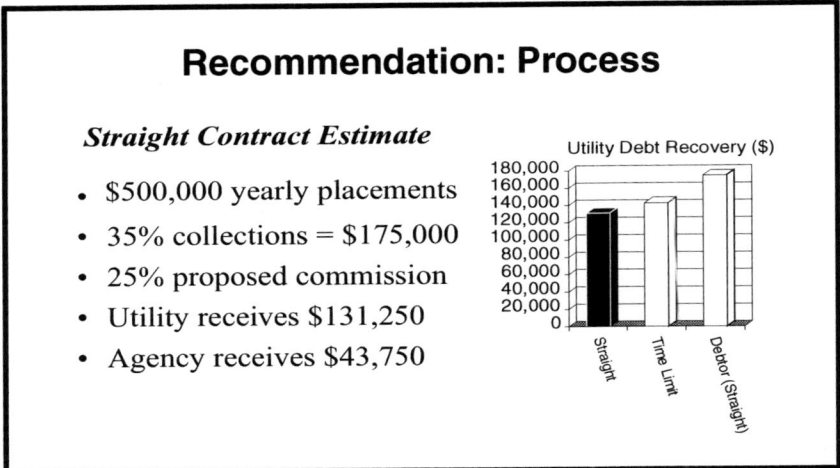

Figure 5-14. Straight Contract Estimate

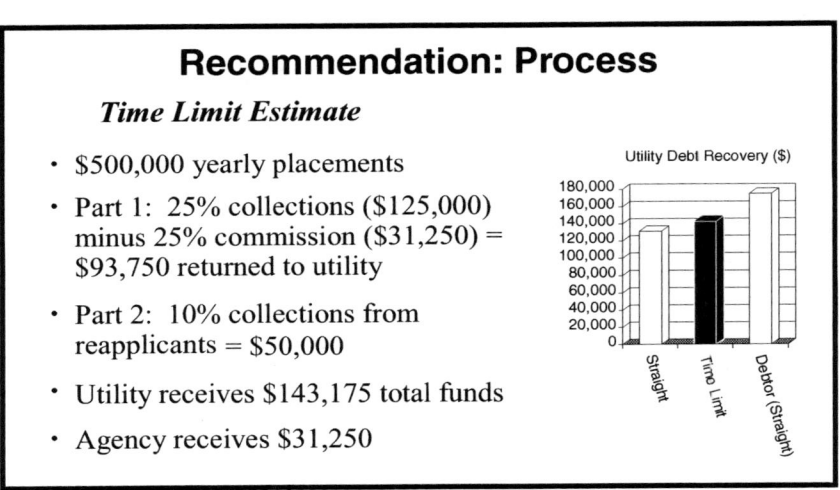

Figure 5-15. Time Limit Estimate

Notes

1. "Call Management System Improves Utility's Collection Results," *Public Utilities Fortnightly* (September 1985).
2. Web conference by National Rural Electric Cooperative Association, April 14, 2004.

DELINQUENT ACCOUNT ACTIONS

> **Recommendation: Process**
>
> *Debtor Pays Commission "Straight" Estimate*
>
> - $500,000 yearly placements
> - 35% collections = $175,000
> - 25% commission paid by debtor
> - Utility receives $175,000
> - Agency receives $43,750
>
> Utility Debt Recovery ($) — bar chart showing Straight, Time Limit, Debtor (Straight)

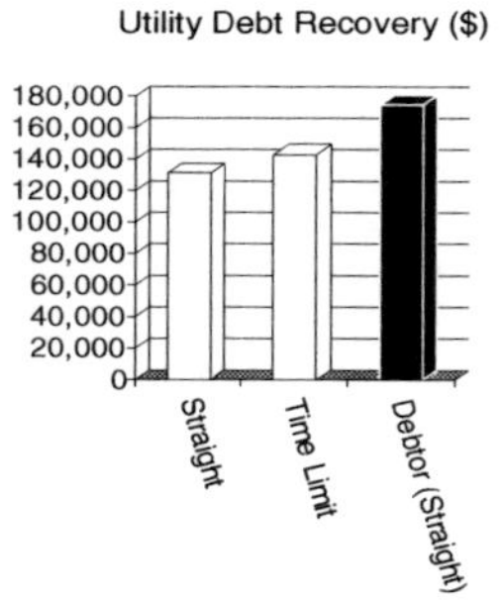

Figure 5-16. Debtor Pays Commission "Straight" Estimate

3. Web conference by National Rural Electric Cooperative Association, June 30, 2004.
4. Steven E. Seger, *The Utility Credit and Collection Manual*, Associated Corporate Consultants, Inc., 1999, 50.
5. *Forbes*, July 6, 1998.
6. Barbara L. Vergetis Lundin, "The Bottom Line: Getting Customers To Pay," *Fortnightly's Energy Customer Management* (Summer 2001):74.
7. David Huff, "Four Key Initiatives for Prepayment Success," *Metering International* (2002 Issue 3).
8. Rachel Dibble, "Customers Take the Power Home—Statistical Data Included," *Transmission & Distribution World* (2000).
9. Fisher, Sheehan, & Colton, "Public Finance and General Economics," *FSC's Law & Economics Insights* (July/August 2001):3.
10. Ken Quesnelle, "Pay-As-You-Go-Power: Treating Electricity as a Commodity," for the North American Power Grid (January 2004):2.

6

Untapped Potentials

The preceding chapters addressed several collections and service management topics that influence local officials/directors, management, and staff alike in ways unique to their positions. Some individuals deal more with policy, and others remain focused on the implementation of policies from an operational perspective. The last four key lessons discussed here deal with unique subject matters that typically do not come up daily. The topics of this chapter include minimizing losses, maximizing revenues, liquidating unproductive assets, and establishing a system for monitoring the impact of credit and collections policies on revenues and customer satisfaction ratings.

LESSON 22: BANKRUPTCY PROCESSING

The financially negative impact of businesses entering bankruptcy is reduced and even eliminated by utilities with prudent long-term credit policies.

Much could be discussed about the activities involved in processing customers' bankruptcy accounts at water, wastewater, electric, and natural gas utilities. Although a detailed, step-by-step review of these activities is beyond the scope of this book, two key issues are worth noting. The worth of this coverage is measured in the amount of monies that utilities lose annually when bankruptcies are filed.

In the short run, just about all utilities incur more bad debt losses from residential accounts than from businesses. In the long run, however, this conclusion is not as strong. Utilities everywhere tell "war stories" about a local plant closing, a healthcare facility going under, or other industries entering default. The challenge in remaining prepared for these large-scale defaults results from the infrequency of their occurrences. Because business defaults happen less frequently than residential incidents, the expectation of their occurring and vigilance to remain prepared fades away.

Three distinct policies designed to insulate revenues from the periodic large-scale losses that result from businesses entering bankruptcy are found at various utilities. The first policy concerns the length of time utilities retain deposits. An example of various lengths of time that business deposits are retained is listed in Table 6-1 for a group of utilities located in Delaware.

The fact that various policies exist reflects the reality that the length of time deposits are retained is basically a policy choice. The utilities that retain deposits as long as the account is active have taken the first step toward securing their revenues over the long term. As discussed in lesson 5, the ultimate objective is to balance the risk presented by high-dollar business accounts with a reasonable level of security while minimizing negative impacts on service opinions. Tools such as surety bonds, irrevocable letters of credit, deposit insurance, and certificates of deposit (see lessons 5 and 6) accomplish these objectives.

Table 6-1. Example of Various Times Business Deposits are Retained

Deposit ($)	Length of Time Retained
250–1,500	As long as account is active
200	As long as account is active
400	As long as account is active
2 months estimate	As long as account is active
2 months estimate	3 years
250 to 3 months estimate	4 years
2½ months estimate	2 years
Highest winter plus highest summer estimates	2 years

Periodically, deposits must be compared to current usage to discover situations in which disparity has materialized. This is the second policy issue that fiscally sound utilities address. As businesses grow, so does their usage and bills. Where gaps are identified, some utilities simply contact business representatives and change the amount of security on hand. Again, where tools such as surety bonds exist, this process is simply administrative in nature.

The third policy matter deals with the use of deposits once a business enters bankruptcy. In court, it has been argued that deposits are assets of a business and, accordingly, should be returned to a trustee for redistribution to all creditors. From an accounting perspective, this makes sense because deposits are listed as assets by businesses and liabilities by utilities on the financial statements of each body. Fortunately, utilities do have a legal basis from which to counter this argument.

> In high-dollar cases, some utilities have successfully obtained court consent to apply deposits by using the legal doctrine of "recoupment." In the case of Norsal Industries, Inc., Long Island Lighting Company (LILCO) effectively utilized this doctrine while defending its application of Norsal's existing $6,395 cash deposits to the debts it owed LILCO. The necessity of relying specifically on recoupment as the authority for applying deposits rests on the fact that what is referred to in common language and legal doctrine as "setoff" is not a legal means for undertaking such action. "[S]etoff of a prepetition debt against a prepetition claim is explicitly stayed by virtue of Section 362(a)(7)." [1]
>
> Contrary to the limitations placed on setoff by the Code as a rationale for applying deposits, no such constraints have been assigned to recoupment. "The justification for the recoupment doctrine is that where the creditor's claim against the debtor arises from the same transaction as the debtor's claim, it is essentially a defense to the debtor's claim against the creditor rather than a mutual obligation, and application for the limitations on setoff in bankruptcy would be inequitable." [2]

The court is stating that the original security deposit and all subsequent bills generated thereafter are part of the same transaction. The transaction was the supply of utility services to customers. They are not separate issues. Therefore, applying deposits to debts is the same as, or similar to, applying routine payments to monthly bills.

LESSON 23: SWEEP ACCOUNTS

Leveraging cash assets provides not-for-profit utilities with an additional revenue stream from which to fund ongoing operations.

The previous key lessons discussed in this book provide a blueprint for crafting policies and operational procedures intended to balance collections performance and service ratings. These lessons work in conjunction with one another, providing information that will help utilities to achieve optimal performance. Once these lessons have been incorporated into a utility's business practices, an additional source of revenue that can be maximized is the sweep account.

Financial institutions offer checking accounts in which excess funds are "swept" into overnight investments daily. After checks and deposits are cleared, the balance is used to draw interest income. Not-for-profit utilities authorized to participate in this option gain revenues that they pour back into operations.

The rate of return fluctuates with the market (i.e., other short-term interest rates) and normally is just below the federal funds rate. Recently, that would be a rate of return of between .5% and 1.5%. The following example illustrates the potential dollars obtainable through sweep accounts.

If a utility's average daily sweep amount is $1,000,000 and if the interest rate the bank is paying on sweep arrangements is 1.5%, the earnings on the account would be $15,000 per year. In practice, a true average daily balance depends on expense levels, collection days, and similar factors. Therefore, utilities that have adopted the most effective new account requirements, current account tasks, and delinquent account actions maintain the highest average daily balances and receive the best earnings.

LESSON 24: SELLING DEBTS

Liquidating uncollected bad debts is a relatively simple and effective business practice for obtaining additional revenues.

Historically, some municipal and cooperative utilities have maintained relationships with two collections agencies. Unpaid accounts have been placed with an agency referred to as a "first placement." The accounts would remain with these agencies for a set period of time, such as 6 months. Afterward, the accounts would be reclaimed from the first agency and placed with the other agency. This action is referred to as a "second placement." Second-placement agencies usually were more productive than other agencies at collecting from especially challenging accounts.

Recalling lesson 21, Collections Contract Terms, debtors pay approximately 75% or more of monies that will be collected within the first 6 months of placement with an agency. Thereafter, debts are "shelved" and no active collections efforts continue until those owing money are denied a credit purchase. To address this reality, as well as to gain additional revenues from unproductive assets (i.e., uncollected debts), some utilities have modified the delinquent account actions flowchart from chapter 1 as depicted in Figure 6-1.

The new collections flowchart offers not-for-profit utilities the opportunity to increase collections with second placements or obtain immediate revenues through the sale of uncollected debts. As with traditional first placements, second placements involve placing debts with an agency. Most collections people in the utility industry know the placement process, but the sale of uncollected debts has not been a common practice for most not-for-profit utility personnel.

In the flowchart, it should be noted that utilities are given the option to either pursue second placements or pursue debt sales, but not both. For reasons that will be explained, it is highly unlikely that debts could be sold for a reasonable value once they have gone through a second placement. Accordingly, utilities are best served by choosing only one option.

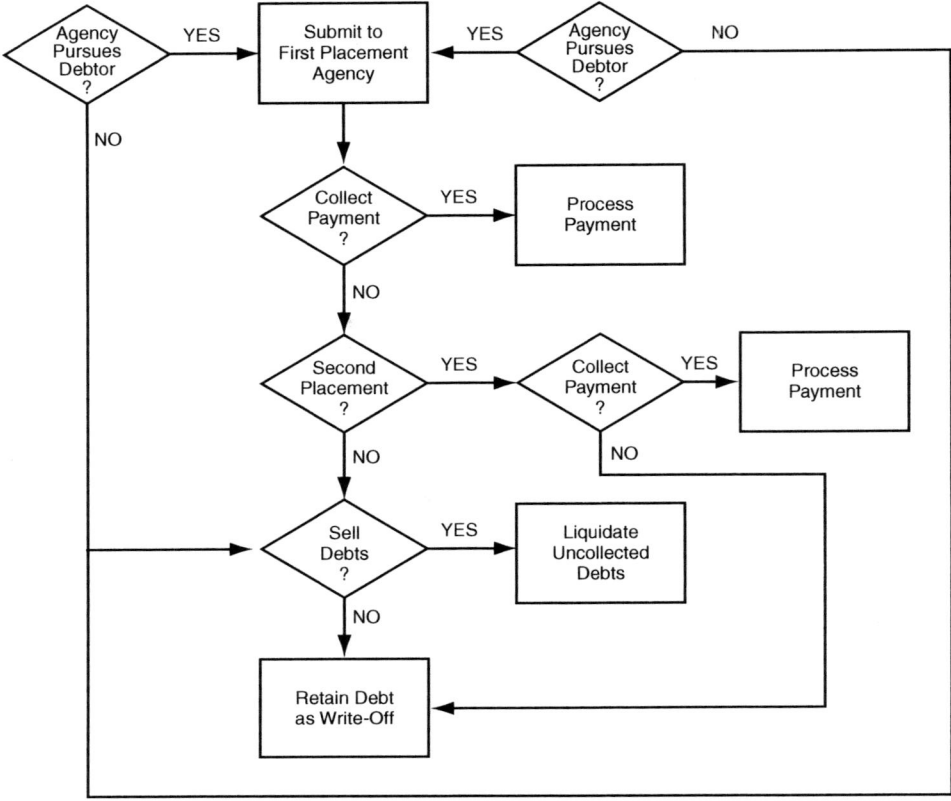

Figure 6-1. Delinquent Account Actions, Modified

The process of selling debts is very much like placing them with an agency for collections. The account information is simply transferred from the utility to the vendor. The only differences with a sale are that a utility receives monies immediately and that the vendor takes ownership of the debts. Utilities that are interested in this option have two opportunities: aggregate backlog and forward-flow sales.

The mountains of bad debt resting in the long-term write-off reports of municipal and cooperative utilities' accounting systems are an aggregate backlog ripe for selling. During the pilot study, an estimation of the aggregate backlog debt was undertaken using information from accounting records. The 10 utilities had approximately $6,238,425 in uncollected bad debts from the previous 5-year period available for selling. The utilities could earn approximately $250,000 if these debts were liquidated for a sale price of 4%. Table 6-2 provides an estimate of the monies that utilities could expect to receive from selling aggregate backlog debts.

Companies that purchase bad debts collect on approximately 6% to 10% of the accounts purchased, depending on qualitative aspects of the debts themselves. These businesses can pay approximately 3% to 5%, therefore, and make a reasonable return on their investments. The key to obtaining the highest price for a portfolio of bad debts is to sell them as soon as possible. Because the likelihood of collections decreases with time, the value of the accounts varies based largely on their age.

A few of the companies that are experienced at valuing utility bad debts and purchasing them include InoVision, Calvary Investments LLC, and Asset Acceptance Capital Corporation. Not-for-profit utilities interested in selling bad debts are encouraged to seek multiple bids for their aggregate backlog portfolios. In addition, utilities can sell their future debts on a forward-flow basis. According to Joel Lewis, president of InoVision, "debts that go uncollected after a set period of time by an agency are sold on a monthly, quarterly, or yearly basis."

Not-for-profit utilities that wish to sell debts on a forward-flow basis must determine the optimal time when accounts should be sold. An investor-owned utility, Commonwealth Edison, provides an example of the age at which it found that selling accounts provided the utility with the highest rate of return.

> Commonwealth Edison, a Chicago utility owned by Exelon, has found that 12 months after charge-off is the optimal point for selling its bad accounts. Up until then the monthly returns on collection efforts outweigh the price it can get from a buyer, said Mark Falcone, director of revenue management for ComEd.[3]

The revenue potential to not-for-profit utilities from the sale of uncollected bad debts varies as to the dollar amount of these debts as well as their age. Table 6-3 is provided as a way for industry personnel to obtain a

BALANCING COLLECTIONS PERFORMANCE AND SERVICE RATINGS

Table 6-2. Backlog Value of Bad Debts

Amount placed for collection per year	100,000	200,000	300,000	400,000	500,000	750,000	1,000,000
70% remains uncollected	70%	70%	70%	70%	70%	70%	70%
Uncollected bad debts per year	70,000	140,000	210,000	280,000	350,000	525,000	700,000
Five years of backlog uncollectibles	5	5	5	5	5	5	5
Total uncollected bad debts	350,000	700,000	1,050,000	1,400,000	1,750,000	2,625,000	3,500,000
Sell bad debts-portfolio value 4% est.	4%	4%	4%	4%	4%	4%	4%
SALES VALUE OF BACKLOG PORTFOLIO	$14,000	$28,000	$42,000	$56,000	$70,000	$105,000	$140,000

Table 6-3. Forward-Flow Value of Bad Debts

Amount placed for collection per year	100,000	200,000	300,000	400,000	500,000	750,000	1,000,000
70% remains uncollected	70%	70%	70%	70%	70%	70%	70%
Uncollected bad debts per year	70,000	140,000	210,000	280,000	350,000	525,000	700,000
Sell bad debts-portfolio value 3% est.	3%	3%	3%	3%	3%	3%	3%
ANNUAL VALUE OF PORTFOLIO	$2,100	$4,200	$6,300	$8,400	$10,500	$15,750	$21,000

quick estimate of the monies they could expect from establishing a forward-flow arrangement with a debt purchasing company.

When considering the sales option, it is important to remember that these debts are the ones that for the most part go uncollected over the long run. Similar to the revenues discussed in lesson 23, Sweep Accounts, it is hard to find a rationale for ignoring free money wherever it is available.

LESSON 25: TRACKING ACCOUNTS RECEIVABLE, COLLECTIONS, AND WRITE-OFFS

Monitoring present collections results is the best way to avoid lagging receivables and heightened losses.

Collectively, the policy and operational matters discussed in this book have a strong impact on the collections performance and service ratings of not-for-profit utilities. Service ratings most commonly are obtained by contracting with third-party companies that gather this information via telephone, mail, or some version of in-person interviews. Ratings are gathered on a monthly basis, on a fixed periodic basis (i.e., yearly), sporadically, or never. Obviously, the ability of local officials and management to monitor the impact of policies and business practices on service ratings is limited to how often these ratings are gathered or, in many cases, whether they are ever gathered.

Whereas gathering service ratings is a choice made at not-for-profit utilities, indicators of collections performance are generated because of financial accounting requirements. Bad debts, uncollected accounts, or write-offs—whatever term is used—is the most prominent figure reported and discussed in the industry. In addition to this financial accounting figure, two managerial accounting items generated to monitor performance are the accounts receivable aging schedule and a collections report.

Accounts Receivable Aging Schedule

A utility's accounts receivable aging schedule primarily indicates the effectiveness of its current account tasks (see chapter 4). The combined effect of billing, pre-shut-off timelines, bad check policy, extensions, and late fees leads to the timeliness of receivables. A June receivables schedule for the pilot study participants is provided in Table 6-4.

The most important item to monitor in an aging schedule with regard to the effectiveness of current account tasks is the current receivables or <30 days category. Table 6-4 provides a good representation of the disparity across the industry in current receivables. Higher percentages indicate greater effectiveness. Based on reviews of collections operations at utilities, average monthly current receivables below 94% at municipals and 90% at cooperatives should be investigated for ineffectiveness, and corrections should be implemented. The average is determined by adding together the <30-day figures from a 1-year period and dividing that total by 12.

Table 6-4. Accounts Receivable Aging Schedule for Pilot Study Participants

Utility	<30 Days	31–60 Days	61–90 Days	>91 Days
1	94.00	2.00	1.00	3.00
2	94.00	4.00	1.00	1.00
3	91.00	7.00	1.70	0.30
4	87.00	5.00	2.00	6.00
5	83.00	16.00	0.70	0.30
6	96.00	2.70	1.00	0.30
7	82.00	6.00	7.00	5.00
8	84.00	6.00	6.00	4.00
9	94.00	4.00	1.99	0.01
10	93.00	5.00	1.00	1.00

Collections Report

A utility's collections report primarily indicates the effectiveness of its delinquent account actions (see chapter 5). A sample collections report modified from a version used by one of the pilot study participants is given in Table 6-5.

All of the items that constitute a utility's delinquent account actions—automatic phone dialers, field collections, etc.—work in tandem, producing a utility's collections effectiveness. The format of the collections report enables utility personnel to monitor the impact of their in-house (i.e., utility personnel) and outsourced (i.e., collection agency) collections efforts. Some utilities may want to add an additional column to track the cost of in-house efforts.

A utility's net write-offs indicate primarily the effectiveness of its new account requirements (see chapter 3). It might appear counterintuitive that net write-offs have more to do with the activation process of an account than with the effectiveness of collections. But the activation process is where the future opportunities and limitations of collections possibilities are created. In short, the thoroughness of information gathered, amount of deposit obtained, and assessment of credit worth on the front end of a utility–customer relationship leads directly to the dollars of net loss experienced once services have ceased. The graph of write-offs for one utility, Figure 6-2, reveals how changes are reviewed over time.

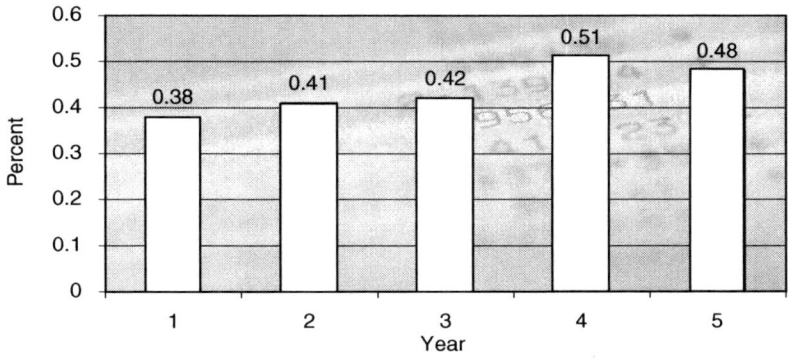

Figure 6-2. Bad Debt as Percent of Revenue

Table 6-5. Sample Collections Report

Gross Year	Utility Personnel Write-Off	Collections Agency Recoveries	Agency Recoveries	Net Commission	Write-Off
1	72,171.66	12,143.30	21,795.60	8,718.24	46,951.00
2	121,467.65	25,299.82	20,857.48	8,342.99	83,653.35
3	92,838.38	23,169.97	26,058.40	10,423.36	54,033.37
4	135,562.27	24,605.98	21,729.08	8,691.63	97,918.85
5	114,949.86	28,114.33	20,850.85	8,340.34	74,325.02
6	140,150.09	40,771.59	18,474.58	7,389.83	88,293.76
7	174,583.61	15,786.88	15,806.05	6,322.42	149,313.10
8	126,243.47	19,598.65	15,652.18	6,260.87	97,253.52
9	250,986.93	33,106.57	18,017.08	7,206.83	207,070.12
10	226,786.53	53,971.40	30,841.95	12,336.78	154,309.96

A review of industry survey data, evaluations of collections operations, and information gathered during the pilot study indicate that net write-offs greater than .15% at municipal utilities and .25% at cooperatives exceed the threshold of what are truly debts that cannot be collected. These types of debts result from situations such as bankruptcies and deaths.

Notes

1. Norsal Industries, Inc., United States Bankruptcy Court for the Eastern District of New York. Case no. 092-71629-21 (1992), quoted in Steven E. Seger, *Utility Credit and Collection Manual*, Associated Corporate Consultants, Inc., 1999, 121.
2. Ibid.
3. David Bogoslaw, "Utilities Make Like Credit Firms by Selling Bad Debt," *Wall Street Journal* (September 25, 2002).

Appendix A

45 Credit and Collection Policies

The process of identifying relevant policies for analysis involved both objective analysis and subjective consideration. Initial questions that provided little or no information were discarded, along with others that offered inconclusive data. After the initial screening of questions and data was complete, 45 policies were selected for statistical analysis. Topical areas of research related to each policy, as well as the policies themselves, are listed in this appendix.

APPLICATION FOR SERVICE

1. The number of methods by which residential customers could apply for service
2. Whether or not residential customers were required to sign an application for service
3. Whether or not residential customers had to provide identification when applying for service
4. Whether or not residential customers had to provide documents (e.g., lease agreement) when applying for service
5. The number of methods by which commercial customers could apply for service
6. Whether or not commercial customers were required to sign an application for service
7. Whether or not commercial customers had to provide identification when applying for service
8. Whether or not commercial customers had to provide documents (e.g., incorporation) when applying for service

APPLICATION FEES AND DEPOSITS

9. The amount of any new fees, other than construction, charged to customers
10. The amount of any fees required for customers to transfer service from one location to another
11. Use of credit scoring to avoid residential security deposits
12. The minimum deposit required of residential property owners
13. The maximum deposit required of residential property owners
14. The minimum deposit required of residential property renters
15. The maximum deposit required of residential property renters
16. Requirement to pay full deposit prior to receiving service
17. Maximum period of time to pay full deposit if not required prior to receiving service
18. Number of available payment options to meet residential security deposit requirements
19. Period of time residential deposits are retained before being applied to accounts or refunded, as applicable
20. Payment of interest on residential deposits
21. Use of credit scoring to avoid commercial security deposits
22. The minimum deposit for commercial accounts
23. The maximum deposit for commercial accounts
24. Number of available payment options to meet commercial security deposit requirements
25. Period of time commercial deposits are retained before being applied to accounts or refunded, as applicable
26. Payment of interest on commercial deposits
27. How often commercial deposits are compared to current usage and additional deposits required, as necessary

BILLING: DISTRIBUTION AND COLLECTION

28. The number of days authorized to pay monthly bills

DELINQUENT PAYMENTS

29. Whether or not customers are notified of delinquent payments
30. Whether or not payment arrangements and/or extensions are available
31. The terms of payment arrangements and/or extensions
32. The amount of late fees
33. The number of days after a due date that late fees are applied to an account

SHUT-OFF

34. Whether or not customers are notified of scheduled shut-offs
35. The number of days shut-off notices gave customers to pay their bills prior to loss of service
36. Whether or not shut-off personnel collect funds in the field
37. Whether separate fees are charged for both shut-off and turn-on

TURN-ON

38. The daytime turn-on fee
39. The amount required to pay prior to a turn-on
40. The amount of additional security deposit required to pay prior to a turn-on

POST–SHUT-OFF COLLECTIONS

41. Whether or not debtors pay commissions owed to collection agencies as opposed to utilities paying them

CREDIT AND COLLECTIONS MISCELLANEOUS

42. Fee charged for non-sufficient funds (NSF) checks
43. Number of days customers are authorized to pay for NSF checks prior to turn-off
44. Number of NSF checks accepted before customers are required to pay by other means
45. Whether utility offers a low-income-customer distribution program

Appendix B

Collections Agency Request for Proposals

I. BACKGROUND

Utility provides water and electric services in state. In FY 200__, Utility had operating revenues of about $___ million. Utility serves approximately _____ residential and business customers.

II. REQUESTED SERVICE

Utility requests proposals for the collection of delinquent Final Bill/Bad Debt customer accounts. Utility is interested in a company whose primary business concerns, professional qualifications, technical competence, and specialized experience indicate ability and willingness to satisfactorily perform this service. In FY 200__, Utility had about $_____ in delinquent accounts available for collections.

Utility intends to electronically submit delinquent accounts to the contractor on a monthly basis. Accounts will be aged approximately 30 days from the date of final billing.

III. MINIMUM REQUIREMENTS

A. Contractor must have a proven track record of at least three years performing services similar to those defined in this Request for Proposals.

B. Contractor's computer system must be able to interface with Utility's communications package. Data will be sent to contractor as a _____ file.

C. Contractor must supply references from at least three previous clients for similar assignments and a description of each engagement.

IV. CONTENTS OF PROPOSALS

Proposals *must* include, at minimum, information pertaining to the items included in this section.

A. References. Contractor is to provide a minimum of three references that include company/agency name, address, contact person's name, and phone number. A brief description of the services performed by the agency as a contractor performing services similar to those described in this request for proposals shall accompany the reference.

B. Notices. Contractor will submit a list of any citations, notices of violation, legal proceedings, or project terminations that any federal, state, or local regulatory agency or department, or corporation or individual has issued to the agency, or any employee of the agency, while that employee was performing work for the agency in the past three years. If there are no violations, contractor shall provide a statement of such.

C. Organization. Contractor will state the (1) number of collectors and (2) average number of accounts assigned to each collector. Also, contractor will provide a description of the agency's organizational set-up and organization flowchart reflecting that the agency can perform the majority of the services with the agency's own forces and equipment under the management of its own organization.

D. Compensation. Contractor will list all fees, commissions, costs, etc., related to the proposed services. A schedule of payments will also be provided.

E. Reports. Contractor *must* provide a SAMPLE REPORT that will be submitted to Utility on at least a monthly basis detailing various matters related to accounts placed with the agency and the status of their collections.

F. Collection procedures.
   a. Contractor will describe its collection procedures from placement to payment reporting. Include in this answer the timing of reports to credit bureaus and customer contacts, types of contacts, and when a file is closed and returned.
   b. Contractor will state the average collection time for its accounts from placement to payment in full during the most recent 12-month period.

G. Term. Contractor will state the duration of the proposal as well as any conditional qualifications for acceptance by Owner.

H. Indemnification. The undersigned vendor, by signing and executing this proposal, certifies and represents to Utility that vendor has not offered, conferred, or agreed to confer any pecuniary benefit, as defined by _____ State Code, or any other thing of value as consideration for the receipt of information or any special treatment of advantage relating to this proposal; the vendor also certifies and represents that the vendor has not offered, conferred, or agreed to confer any pecuniary benefit or other thing of value as consideration for the recipient's decision, opinion, recommendation, vote, or other exercise of discretion concerning this proposal. The vendor certifies and represents that vendor has neither coerced nor attempted to influence the exercise of discretion by any officer, trustee, agent, or employee of Utility concerning this proposal on the basis of any consideration not authorized by law; the vendor also certifies and represents that vendor has not received any information not available to other vendors so as to give the undersigned a preferential advantage with respect to this proposal; the vendor further certifies and represents that vendor has not violated any state, federal, or local law, regulation, or ordinance relating to bribery, improper influence, collusion, or the like, and that vendor will not in the future offer, confer, or agree to confer any pecuniary benefit or other thing of value of any officer, trustee, agent, or employee of Utility in return for the person having exercised their person's official discretion, power, or duty with respect to this proposal. The vendor certifies and represents that it has not now and will not in the future offer, confer, or agree to confer a pecuniary benefit or other thing of value to any officer, trustee, agent, or employee of Utility in connection with information regarding this proposal, the submission of this proposal, the award of this proposal, or the performance, delivery, or sale pursuant to this proposal.

I. Rights to extend submission date. Utility reserves the right to reject or re-solicit if only one response or none is received by the "submission date" or extend the submission date by an additional two (2) weeks.

V. QUESTIONS REGARDING THE RFP

E-mail any technical issue and specification questions pertaining to this Request for Proposals ("RFP") to (name) on or before (date). Specifically reference the section of the proposal in question. All questions must be submitted via e-mail. Questions and answers will be distributed to all suppliers solicited in order to avoid any unfair advantage. These guidelines for communication have been established to ensure a fair and equitable evaluation process for all respondents. Any attempt to bypass the above lines of communication may be perceived as establishing an unfair or biased process and could lead to disqualification as a potential service provider.

VI. PROPOSAL SUBMISSION

Submit four complete copies of the proposal along with one signed copy of this Request for Proposals to the following address on or before 5:00 P.M., __DT, (date).

APPENDIX B

Name _____

Address_____

VII. REJECTION OF PROPOSALS

Utility reserves the right to reject any and all proposals and at its sole discretion may solicit for proposals using the same or a different document.

VIII. AWARD OF SERVICES

The award of services, if they are awarded, will be made by Utility. The determination of services to be provided by the selected contractor shall be at the sole discretion of Utility. Should the successful contractor and Utility fail to come to an agreement, Utility may at its sole discretion award services to the next most qualified, responsive, and responsible contractor.

The contractor to whom the contract is awarded shall be required to negotiate and enter into a written commercial terms contract with Utility in a final form approved by legal counsel of Utility.

I have read and understand this Request for Proposals.

_____

Signature

_____

Date

_____

Type Name

_____

Organization

Index

Note: *f.* indicates figure; *t.* indicates table.

Accounts receivable aging schedules, 94, 94*t*.
"Analyse-It General Statistical Module" software, 8
Application fees and deposits, x, 22
   credit management tools, 26, 27*f*., 28*f*., 29*f*.
   deposit insurance, 27*f*.
   disposition in bankruptcy proceedings, 87–88
   effect on service ratings, 22, 23–24, 25*f*.
   effect on write-offs, 22–23, 24*t*.
   increasing as usage increases, 87
   irrevocable letter of credit, 29*f*.
   minimum deposits and years kept, 24–26, 25*t*.
   policies selected for statistical analysis, 100
   retention time and protection against customer bankruptcy, 86, 86*t*.
   sample dollar amounts and acceptable forms of deposits, 22, 24*t*.
   utility service guaranty bond, 28*f*.
Application for service, x
   effect on service ratings, 21–22, 22*t*.
   effect on write-offs, 21
   policies selected for statistical analysis, 99
   samples, 23*f*.
Automatic phone-dialer contacts, xi, 51–53, 53*f*.

Bad checks, xi, 41
   policies and fees, 41–42, 42*t*., 43*f*.
Bankruptcy processing, xii, 85–88
Billing frequency, x, 35–38, 36*t*., 37*f*., 38*f*.
   policies selected for statistical analysis, 101
   and presentment methods, 35

Certificates of deposit, x, 26–30, 30*f*.
Collection priority, x, 6, 11
   effect on current accounts receivable, 12*t*., 14*t*.
   effect on satisfaction ratings, 13*t*.
   effect on write-offs, 12*t*.
   and Wilcoxon signed-ranks test, 13*t*.
Collections agency request for proposals, xii, 74–75
   case study, 75–79, 76*t*., 77*t*.–78*t*.
   sample, 103–107
Collections contract terms, xii, 76
   case study, 81–82
   commissions, 79–81, 80*t*.
   debtor paying commission, 79, 81, 83*f*
   duration of contract, 80–81, 82*f*.
   fees, 80
   policies selected for statistical analysis, 101
   straight contract, 81, 82*f*.
Collections reports, 95–97, 95*f*., 96*f*.
CP. *See* Collection priority
Credit databases, x, 31–33
Credit scoring, x, 30–31, 32*t*.
Cross-tabulated contingency tables, 8
Cross-tabulation analysis, 20
CSP. *See* Customer service priority
Current account tasks, 9, 35.

See also Bad checks, Billing frequency, Late fees, Multiple active timelines, Payment extensions and arrangements

Customer service priority, x, 6, 15
    effect on current accounts receivable, 16*t*., 17
    effect on satisfaction rating, 16*t*., 17
    effect on write-offs, 15*t*., 17

Deconstructing bad debts, xii, 63
    active timeline, 66, 66*f*., 66–69
    case study, 63–71
    delinquent accounts receivable, 65, 65*t*.
    deposit threshold, 71–72
    factors contributing to losses, 67, 67*f*.
    inactive timeline, 66*f*., 69–70, 69*f*., 70*f*.
    miscellaneous contributing factors, 71
    revenues vs. bad debts, 65–66, 65*t*.

Delinquent account actions, 9, 10*f*., 51, 90*f*.
    policies selected for statistical analysis, 101
    See also Automatic phone-dialer contacts, Collections agency request for proposals, Collections contract terms, Deconstructing bad debts, Disconnection/shut-off, Field collections, In-house versus outsourcing collections, Selling debts, Technologies for disconnection/reconnection

Deposits. *See* Application fees and deposits, Certificates of deposit

Disconnection/shut-off, xi, 56–59
    door hangers, 56–59, 57*t*., 58*f*.
    mailed shut-off notices, 59, 59*f*.
    policies selected for statistical analysis, 101

Equifax
    and credit databases, 33
    and credit-scoring services, 31

Experian, 31

Field collections, xi, 53–56
    and theft or violence, 54, 55*f*.

Francisca Osilek et al. v. Commonwealth Utilities Corporation (CUC) et al., 57

In-house versus outsourcing collections, xii, 71
    collections flowchart, 72, 72*f*.
    credit bureaus, 74
    in-house results, 71–73
    locating debtors, 73–74
    procurement methods, 74
    skip tracing services, 73

Inactive timeline, x, 17
    and Wilcoxon signed-ranks test, 19*t*.
    and write-offs, 17, 18*f*., 18*t*.

Key lessons, x–xii

Late fees, xi, 46–47, 49
    inadequate, 48
    nonexistent, 47–48

Lessons, x–xii

Low-income-customer policies selected for statistical analysis, 101

Meter reading, 37

Multiple active timelines, xi, 39
    "friendly" and "termination" categories, 39, 40*t*., 41*t*.
    rating systems for late payers, 39–41, 39*t*., 40*t*., 41*t*.

New account requirements, 8–9, 21. *See also* Application fees and deposits, Application for service, Certificates of deposit, Credit databases, Credit scoring

Non-sufficient funds policies selected for statistical analysis, 101

ONLINE, 33

Payment extensions and arrangements, xi, 43–45, 44*f*., 45*t*.
    arrangement defined, 43
    extension defined, 43
    sample application form, 45, 46*f*.

Pilot study of policies and results, 1
    answers defined as CP or CSP, 6, 7*f*.

data collection, 3, 5
data preparation, 5–6, 6f.
design of study, 2–3
hypotheses, 7–8
key policies as independent variables, 6
number of customers of participating facilities, 4, 4f.
participants, 3–4
performance indicators as dependent variables, 6
testing, 8
topical areas, 2–3
Predictive dialer systems. *See* Automatic phone-dialer contacts
Prioritizing collections. *See* Collection priority
Prioritizing service. *See* Customer service priority
p-value, 13–14

RoadMAPS, 37
RouteMAPS, 37

SchlumbergerSema, 37

Selling debts, xii, 89–93, 90f., 92t.
representative buyers, 91
Sweep accounts, xii, 88

Technologies for disconnection/reconnection, xi, 59–63
and costs, 59–60, 61t.
limiters, 62
prepay meters, 60–62
remote devices, 62–63, 64t.
Tennessee Valley Authority, 3–4
Tennessee Valley Public Power Association, 4
Tracking accounts receivable, collections, and write-offs, xii, 93
accounts receivable aging schedule, 94, 94t.
collections report, 95–97, 95f., 96f.
TransUnion, 31
Turn-on policies selected for statistical analysis, 101
25 key lessons, x–xii

W statistic, 14
Wilcoxon signed-ranks test, 8, 13

ABOUT THE AUTHOR

Steven E. Seger is a Principal with Seger Consulting Group, Inc. (SCG), a management consulting and training company that has been providing credit and collection services to utilities and government agencies since 1973. Steve has a Bachelor of Finance and a Masters in Business Administration.

His writing projects include industry manuals and articles, and he has served as a publisher. A few examples are:

- *The Utility Credit & Collection Manual*
- American Gas Association's *The Service Connection* (former publisher)
- "Assessing The Quality of Service Provided by Public Power Systems" (report for the American Public Power Association)
- Numerous articles for industry publications

Steve routinely conducts training for utility personnel from the United States and abroad. In addition, he works with individual utility credit and collections personnel to determine the cost-effectiveness of their programs. Based on this information, recommendations for reducing bad debts has saved utilities and government agencies millions of dollars in previously lost revenue.